Reliability Modelling and Optimization of Warm Standby Systems

Rui Peng · Qingqing Zhai · Jun Yang

Reliability Modelling and Optimization of Warm Standby Systems

国防工業出版社
National Defense Industry Press

Springer

Rui Peng
School of Economics and Management
Beijing University of Technology
Beijing, China

Qingqing Zhai
School of Management
Shanghai University
Shanghai, China

Jun Yang
School of Reliability and Systems Engineering
Beihang University
Beijing, China

Translated by

Rui Peng
School of Economics and Management
Beijing University of Technology
Beijing, China

Qingqing Zhai
School of Management
Shanghai University
Shanghai, China

Jun Yang
School of Reliability and Systems Engineering
Beihang University
Beijing, China

Heping Jia
School of Economics and Management
North China Electric Power University
Beijing, China

ISBN 978-981-16-1794-2 ISBN 978-981-16-1792-8 (eBook)
https://doi.org/10.1007/978-981-16-1792-8

Jointly published with National Defense Industry Press.
The print edition is not for sale in China (Mainland). Customers from China (Mainland) please order the print book from: National Defense Industry Press.

This Springer imprint is published by the registered company Springer Nature Singapore Pte Ltd.
The registered company address is: 152 Beach Road, #21-01/04 Gateway East, Singapore 189721, Singapore

Preface

Since the 1950s of the twentieth century, reliability as a special subject was born. The development of reliability has always been closely combined with the scale and modernization of production. With the improvement of production technology, the reliability of single product is constantly improved. However, the increasingly complex system brings challenges to the design of system reliability. As a basic method of reliability design, redundancy and backup design is always an effective means to ensure and improve system reliability. The warm backup design combines the characteristics of cold backup and hot backup and has the advantages of both. In addition, from the perspective of reliability model, warm standby is a general model including cold standby and hot standby. Studying the reliability characteristics of warm standby has a guiding role in the reliability research of cold standby and hot standby system.

The main content of this book summarizes the research results of Rui Peng, Ph.D. supervisor of Beijing University of Technology, Qingqing Zhai, Associate Professor of Shanghai University, Jun Yang, Professor of Beijing University of Aeronautics and Astronautics, and their collaborators in this field. The reliability modeling and optimization problems of general warm standby system are systematically studied based on probability method. This book mainly uses the method of decision diagram for system reliability modeling, which is not limited to the simple warm standby system with one or two warm standby components, or it is necessary to assume that the life of warm standby components follows exponential distribution, which enriches and develops the reliability modeling and optimization methods of current warm standby systems. The Chinese version of this book belongs to the "new series of reliability books" and has been listed in the "13th five year" key publishing plan of the General Administration of press and publication. The copyright of the Chinese version belongs to the National Defense Industry Press and is supported by the national defense science and technology book publishing fund of the equipment development department of the Central Military Commission.

The structure of the book is as follows. The Chap. 1 summarizes the background knowledge of warm standby and the current research status. The Chap. 2 introduces the working mechanism of warm standby, the common structure of standby system, the fault coverage and the decision diagram method used in this book. In the

Chap. 3, the method of reliability modeling for the most common k out of n warm standby system is given by using decision diagram technique. In the Chap. 4, the reliability modeling method of general warm standby system based on multi value decision diagram is given from the perspective of failure sequence in the system. In Chap. 5, the reliability modeling of warm standby system with fault coverage and switchover failure is further considered. In the Chap. 6, we study the optimal order of unit work in general warm backup system. In Chap. 7, the reliability modeling method of the multi state warm standby system is given. The Chap. 8 discusses the reliability modeling method of warm standby system with common bus structure. In Chap. 9, the reliability modeling method of warm standby system with phased mission requirements is discussed. Chapter 10 summarizes the whole book and discusses the general modeling method of warm standby system in other structures. In the book, the Chaps. 1 and 2 are written by Dr. Qingqing Zhai; the Chaps. 3, 4, 6, 8 and 9 are jointly written by Prof. Rui Peng and Dr. Qingqing Zhai, with the assistance of Dr. Heping Jia From North China Electric Power University; the Chaps. 5, 7, and 10 are written by Prof. Rui Peng. Professor Jun Yang is mainly responsible for the inspection of the whole book and overall grasp the quality of the book.

The research results of this book can be applied to the design and production of aviation equipment, wireless sensor system, power system and other equipment systems with high reliability and long life requirements, which has important and practical reference significance for guiding the reliability design, evaluation and optimal configuration of related equipment or system.

Beijing, China — Rui Peng
Shanghai, China — Qingqing Zhai
Beijing, China — Jun Yang

About This Book

This book systematically studies the reliability modeling method of warm standby system with different structures and solves the reliability modeling problem of existing warm standby system, which has important and practical guiding significance for reliability analysis, evaluation and optimization design of warm standby system. The model proposed in this book is applicable to the system with any number of warm standby components and is applicable to the case that the failure time of system components follows arbitrary distribution. The method used in this book is mainly based on the traditional binary decision diagram and multi value decision diagram. According to the characteristics of warm standby system, this method has great reference value for the reliability research of other complex systems. The readers of this book are researchers, engineers and technicians engaged in reliability research and work as well as graduate students in school.

Contents

Chapter 1
Introduction

Reliability is an important quality characteristic. With the development of modern society and economy, people put forward higher and higher requirements for product reliability. Xuesen Qian said that "reliability is designed, produced and managed", and reliability first depends on product design. As a system, the reliability level of a product depends largely on the reliability of the units in the system. Obviously, when the assembly and testing of the products are the same, the higher the reliability level of the component, the better the reliability of the corresponding products. However, it is difficult to improve the reliability of the system simply by improving the reliability of the component. In this case, the redundant design can be considered, and the components with limited reliability level can be used to construct products meeting the requirements of high reliability.

1.1 Redundancy and Standby

In the field of engineering, redundancy refers to the design feature that the product uses more than the minimum necessary quantity of equipment to complete the specified function so that it can still complete the same specified function when a set of equipment fails. The idea of redundancy design is to repeatedly configure some units, components or functions, and the fundamental purpose is to improve the reliability of products. Due to the existence of redundancy, the same function can be realized by multiple components, and the probability of multiple components failure at the same time is usually much less than that of single component, so the reliability level of products can be improved. For example, the fly by wire flight control system used in modern military or civil aircraft usually contains multiple channels, and also includes a set of redundant hydraulic control system. In this way, when a certain channel fails, the aircraft can still be controlled. For another example, the server system used by large-scale websites contains a large number of servers (as shown in Fig. 1.1), including a certain proportion of redundant servers, so as to ensure that

R. Peng et al., *Reliability Modelling and Optimization of Warm Standby Systems*,
https://doi.org/10.1007/978-981-16-1792-8_1

Fig. 1.1 Google data center in Oklahoma, USA, with a large number of servers arranged in order (*source* Google data centers)

the system can still maintain the normal operation of the website in case of server failure, maintenance or upgrade.

According to the different working mechanism, redundancy design can be divided into active redundancy and standby redundancy. In work redundancy, all units of the redundant system work in parallel, all units complete the same function, and the same work can be completed by any component or distributed among different components. When a component fails, it will no longer participate in the work of the system, and the remaining units can still ensure the normal operation of the system. All units work at the same time, complete the same function and have similar life history. In work redundancy, there is no switching action between standby state and normal working state in work component, so work redundancy is also called static redundancy technology. For example, the brake system of a modern car usually contains multiple brake pads. Due to the wear and aging of brake pads, their reliability is gradually reduced. However, redundant brake pads ensure that the vehicle can still work safely when individual brake pads fail (as shown in Fig. 1.2a). Different from working redundancy, in standby redundancy, the component in the standby state does not directly participate in the work of the system, but only after the normal working

(a) Red brake pads of F1 racing car which can brake the car from 200km / h to full standstill in 2.9 seconds (working redundancy)

(b) Off road vehicles with poor working conditions are usually equipped with spare tires (backup redundancy)

Fig. 1.2 Common redundancy cases in life

component fails, the standby component switches to the normal working state to replace the failed component. The spare tire of a car is the most common example of standby redundancy (Fig. 1.2b). In computer, server, storage system, power network and other areas with high reliability requirements, the design of standby redundancy will be used in the system design. In these systems, the detection of failed component and the switching of standby component are usually realized automatically. Compared with working redundancy, standby redundancy has a series of detection, switching and recovery operations, so it is also called dynamic redundancy technology.

According to the different mechanisms, the standby design can be further divided into hot standby, cold standby and warm standby. In the hot standby system, the standby component runs simultaneously with the normal working component but does not participate in the work and output of the system. For example, in the common "one main and one standby" hot standby server system, two identical servers run simultaneously and backup each other. At any time, only one machine serves as the main work server, while the other machine keeps the machine on and running as a hot standby, receiving and processing data as the main server. The backup server works synchronously with the primary server, and the working status is similar, but it does not provide services to the client (no data is returned). When the primary server fails, the hot standby server will automatically detect the failure and take over the task of the primary server. In the cold standby state, the standby component does not work, so the ideal cold standby component will not degrade or fail, and the length of standby time has no effect on the subsequent normal service life. Taking the server system as an example, cold standby means that the machine is shut down or dormant. Warm standby, as the name suggests, is an intermediate state between cold standby and hot standby. In the warm standby state, the unit also bears a certain working stress, which may lead to failure. However, compared with the normal working stress, the working stress in warm standby state is usually milder, which makes the failure rate of warm standby state lower than that of normal working state. Accordingly, the life distribution of components in warm standby state is usually different from that in working state. Taking the dual server system as an example, the standby component only periodically copies and backs up the data from the primary server in the warm standby system, so its work intensity is lower than that in the hot standby state, and it is not easy to fail.

From the actual situation, the component in the standby state is usually not only completely free from failure but also different from the normal working stress. Therefore, warm standby can better describe the working state of standby products. From the perspective of reliability model, warm standby can be regarded as the extension of cold standby and hot standby: when the stress under warm standby state approaches to that under normal working condition, and the corresponding failure law approaches to failure law under working state, warm standby tends to hot standby; when the stress under warm standby state is very small and the unit basically does not fail under warm standby state, warm standby tends to be cold standby. Therefore, warm standby model is a more general standby model. Therefore, whether from the perspective of theoretical research or from the perspective of practical application, the warm standby model is worthy of in-depth study. On the one hand, as a generalization model of cold standby and hot standby, the research results of warm standby model can be directly applied to the reliability analysis of cold standby and hot standby systems. On the other hand, compared with hot standby or cold standby, warm standby model is more flexible and can model the actual standby system more accurately.

1.2 Development of Standby System Reliability Modeling

Standby design has been widely used in practice, and its reliability has also been fully valued and studied. As mentioned above, the standby system can be divided into cold standby, hot standby and warm standby. Whether from the actual technical implementation or from the difficulty of reliability modeling, cold standby or hot standby is simpler than warm standby. In the existing literature related to the reliability of standby systems, the early researches were mostly about these two types of standby systems. Before introducing the research of warm standby system, we will first briefly introduce the related research of cold standby and hot standby. Before entering the topic, two different objects considered in current reliability research are explained: repairable system and non-repairable system.

In the reliability research, whether the product can be repaired or not determines the technical methods involved in the research. When the product is repairable, it can still be restored to operation through maintenance and replacement of parts after failure so that the product states can transit between the two states of "intact" and "failure". It is noted that reliability usually refers to the ability of a product to remain in good condition for a certain period of time; when the product state can transit between two states, the total time of the product in good condition can be infinite. In order to compare the reliability level of different products, we can consider the proportion of the time that the product remains in the "good" state in the total service life, or the probability that the product is in the "good" state at a certain time. This involves the concept of availability. In fact, in many researches on the reliability of repairable system, people mainly study the availability of products. For the repairable system, the availability research needs to model the transition of products in different states. In this case, the research object is the set of possible states of the product, namely state space. In order to obtain the exact number of elements in the existing system, for example, the finite number of elements in the system can be considered. In this case, the state space can be obtained by enumeration (or induction), and the key point of system availability analysis is to model the dynamic process of "transition" between different states. For a non-repairable system, once the component in the system fails, it will be in the failure state permanently (at least within the time range we consider), and the state transition is unidirectional. At this time, the transition between different states of the system is much simpler: only the transition from "good" to "failure" is needed. When the failure of each unit is independent of other units, the time when it is in "good" state depends on its own reliability level. At this time, because the transition between different states of the system is clear and definite, the focus of system reliability modeling is on how to obtain the state space corresponding to system failure (normal), that is, to find the so-called cut set (path set). Therefore, although the objects of reliability research are the same for repairable or non-repairable systems, the focus of attention is different, so the solution approaches will be different.

Back to the reliability research of cold standby system, the repairable and non-repairable systems are discussed respectively. As mentioned above, the reliability modeling tool of repairable cold standby system is mainly state space method. Most of the previous studies have considered systems with fewer elements, such as two-element systems. Singh and Agrafiotis [1] studied the reliability of a repairable cold standby system with two identical units using the regeneration point technique. Kumar et al. [2] regarded a kind of carbon recovery device as a cold backup system and solved its reliability by state space method. Pandey et al. [3] studied the reliability of a cold standby textile device using the state space method. Mokaddis et al. [4] studied a repairable cold standby system with two different types of units using semi Markov technique. Wang and Zhang [5] analyzed the availability of a two-unit repairable cold standby system by using state space method considering the priority of component use and maintenance. Tang and Liu [6], Yu et al. [7] and Liang et al. [8] considered the vacation of maintenance workers and studied the reliability of two-unit repairable cold backup system. For the non-repairable cold standby system, the focus of the research is on finding the state set corresponding to the system fault. At this time, there are various technologies used, but the basic principle is combination method. Azaron et al. [9] proposed a network graph method to obtain the system failure state set for cold standby system with exponential distribution for component failure time and solved the system reliability by Markov method. Azaron et al. [10] used the network graph to solve the system reliability with cold standby and components subject to Erlang distribution. Since the calculation of reliability of general cold standby system involves integration, it is difficult to calculate. Wang et al. [11] proposed an approximate calculation method of system reliability based on central limit theorem for 1-out-of-N and K-out-of-n cold standby systems, including independent identically distributed units. When the unit life distribution in the system is not independent, Eryilmaz and Tank [12] used copulas method to study the reliability of a three-unit cold standby system. In conclusion, for the research of cold standby system, the reliability of the system is analyzed according to the state space method and solved by Markov method.

If the reliability of switching is not considered, the system reliability model of hot standby is the same as that of common working redundancy [13]. The common structure of work redundancy is k-out-of-n structure, that is, k-out-of-n units can ensure the normal operation of the system. The related research of this kind of system occupies a large proportion in the field of system reliability modeling. The main reasons can be summarized as two points. First, this kind of system is more common in practice, and its reliability research is of great significance; second, the reliability modeling and analysis of this kind of system involves more complex combination problems (compared with series parallel system), and its reliability analysis is more research-oriented. For example, in Kuo and Zuo's "Optimal Reliability Modeling: Principles and Applications" [13], there are four chapters related to this kind of system. In recent studies, some scholars have considered the more general reliability modeling of weighted k-out-of-n hot standby system [14–17]. Huang et al. [18], Tian et al. [19], and Li and Zuo [16] further considered the reliability modeling of multi-state k-out-of-n system and multi-state weighted k-out-of-n system. Coit et al. [20] studied a k-out-of-n system in the case of component partnership and gave the reliability model of the system. In conclusion, for the general simple hot standby system, since there is no difference between the backup state and the normal working state (for the unit), the solution of the system reliability is more direct, so researchers pay more attention to the availability when the system is repairable or the optimal allocation of spare parts when it is not repairable.

1.2.1 Reliability Modeling of Repairable Warm Standby System

The reliability modeling of warm standby system is also divided into repairable and non-repairable. In the existing reliability modeling research of repairable warm standby system, most of them assume that the life and maintenance time of the unit in normal operation and warm standby state obey exponential distribution, and then the Markov analysis method is used to solve the problem [21–35]. A few researchers consider the non-exponential distribution of random variables and use semi Markov technology to analyze the system reliability [36–41].

When the component life and maintenance time both obey exponential distribution, Markov method can be used to analyze the system reliability. In Wang et al. [22, 24, 27, 28, 32, 33], Ke et al. [31, 34, 42] and Jain et al. [43], a series of researches on k-out-of-n (M + n) repairable warm backup system including m work units and N warm backup components (all components are equally distributed) is studied. The reliability and availability of the system are analyzed by means of Markov method considering such factors as component maintenance waiting and exiting, unreliable maintenance and maintenance pressure. Zhang et al. [26] studied a repairable warm standby system with k-out-of- (M + n) containing two types of components. Assuming that the unit life and maintenance time are exponential distribution, the reliability and availability indexes of the system are solved by Markov method. Wang and Kuo [21] studied the reliability of repairable hybrid standby systems with four different system configurations, in which the component life and maintenance time are exponentially distributed. Yuan and Meng [30] considered a repairable warm standby system with two different exponential components (one work component and one warm standby component) and analyzed the system availability using Markov process and Laplace transform. Hu and Chen [44], Liu et al. [45], and Yang and Zhang [46] analyzed and solved the system reliability by using the state space method for the repairable warm standby system containing two different exponential components.

When the life distribution or maintenance time of a component obeys arbitrary distribution, the state space method is still the main tool for system availability (reliability) modeling, and semi Markov analysis methods may be used. Mokaddis and Tawfek [39] consider a two-unit repairable warm standby system. Assuming that the life distribution of the unit in the working and warm standby states is arbitrary, the system availability is analyzed by using semi Markov technique. Mahmoud and Esmail [36] also consider a two-unit repairable warm standby system. Assuming that the component life follows an exponential distribution, but the maintenance time is a general distribution, the system availability is analyzed by using the regeneration point technique in Markov renewal process. Hsu et al. [38] studied the availability of a repairable warm standby system, including the two work components, a warm standby component and a maintenance equipment. It is assumed that the component life follows the exponential distribution of independent and identically distributed, and the maintenance time obeys the arbitrary distribution, and analyzes the system by using the state space method. Pérez-Ocón and Montoro-Cazorla [40] analyzed the availability of 1-out-of-N repairable warm backup system with phase type distribution of working life and maintenance time of a unit. El-Damcese [41] extended the research of Zhang et al. [26] to the case that component life and maintenance time follow arbitrary distribution, and the common cause of failure was considered. In conclusion, state space analysis is the main method to analyze the reliability and availability of the repairable warm standby system, and the Markov-related technology is the main method to solve the problem.

1.2.2 Reliability Modeling of Non-repairable Warm Standby System

There are few researches on the reliability of non-repairable warm backup system. Li [47] proposed a reliability model for the 1-out-of-N warm standby system with independent identically distributed index units and considering the unreliability of switching. She and Pecht [48], Li et al. [49] and Amari et al. [50] studied the reliability of k-out-of-N backup system with independent identically distributed exponential components and gave the analytical expression of the system reliability. Amari and Dill [51] analyzed the reliability of a hybrid standby system with independent and identically distributed component life which can obey arbitrary distribution. For the k-out-of-N warm standby system with arbitrary component life distribution, Tannous et al. [52] proposed a system reliability calculation method based on sequential binary decision diagram. Although this method is suitable for reliability analysis of K-out-of-N warm backup, it is not a formalized method. Tannous and Xing [53] proposed a method to approximate the system reliability by using the central limit theorem for 1-out-of-N warm backup system, but the effect of the approximate method is not ideal. Levitin and Amari [54] solved the reliability of k-out-of-n warm backup system by using universal generating function (UGF) by discretizing time. Some scholars have analyzed the reliability of some special warm standby systems. Cha et al. [55] and Li et al. [56] studied the reliability of a two-component (one main and one standby) warm standby system. Papageorgiou and Kokolakis [57] studied the reliability of a system with two work components and (n-2) warm backup units and derived the iterative formula of system reliability. Eryilmaz [58] considered the case of a k-out-of-n system with independent and identically distributed components which is additionally equipped with a warm standby unit, and the analytical expression of system reliability function is given. Generally speaking, there is no unified and effective method to model and calculate the system reliability for general non-repairable warm standby system.

1.2.3 Reliability Optimization Design of Standby System

Reliability optimization design of standby system has always been an important research content in the field of reliability [59], which can be divided into two categories according to the research angle. The first kind of problem usually involves design elements such as system reliability (availability), system configuration cost, system total weight and system volume. Taking some (some) elements as constraints (such as system total cost), some (some) elements as objective function (such as system reliability) and standby type and quantity as design variables, a new optimization problem can be proposed and solved. Coit [60] discussed the optimal design problem of non-repairable series parallel cold standby system under imperfect switching. Assuming that the life distribution of components follows Erlang distribution, the optimal allocation of standby components to maximize the system reliability at a given time under cost and weight constraints was constructed, and the optimization problem was transformed into an equivalent 0–1 programming problem to solve. Coit [61] further considered the optimal allocation of standby components for series parallel systems, including cold standby and hot standby, and constructed an optimization problem to maximize system reliability under cost and weight constraints. The problem was also transformed into an equivalent 0–1 programming problem and solved. Coit [62] takes the uncertainty of parameter estimation into account and studies the optimal allocation of standby components in series parallel system with the estimation variance of system reliability as the optimization objective. Boddu and Xing [63] and Boddu and Xing [64] consider a series system consisting of several k_i-out-of-n_i subsystems, in which each subsystem can be configured with mixed cold standby or hot standby. The optimization problem of maximizing system reliability at a certain time under the constraint of configuration cost is constructed, and the optimal standby components allocation is solved by genetic algorithm. For the repairable system, Yu et al. [65] took a two-unit (one for work and one cold backup) repairable system as the object, constructed an optimization problem with the system availability as the constraint, the average repair time and the average time between failures of the component as the design variables, and the total cost as the optimization objective. Some scholars, such as Huang et al. [66], Azaron et al. [67], Safari [68], Ardakan et al. [69], considered the multi-objective spare parts optimization problem, and solved the problem based on genetic algorithm. For general systems with warm standby designs, Amari and Dill [70] considered a series system consisting of several k_i-out-of-n_i subsystems and studied the optimization problem of maximizing the system reliability under the constraints of cost, weight and volume. Tannous [71], respectively, used integer programming and genetic algorithm to solve the optimal allocation of standby components of series parallel warm standby system under the constraints of cost and total spare parts.

The second problem is to consider how to maximize the reliability index of the system by reasonably configuring the location or working order of the standby components given the available components in the system design. In this kind of problem, the standby component configuration in different locations will lead to different system reliability. Since the available standby is given, the cost, weight and other factors are not considered, but the optimization of system reliability index is only considered. The related research often involves the stochastic scheduling problem, which can be generally obtained by analytical method [72]. Romera et al. [73] studied the hot standby configuration of two units in a two-unit series system and a k-out-of-n system. Li and Hu [74] studied the configuration of hot standby unit in series parallel system under various conditions. Hu and Wang [75], respectively, studied the optimal allocation of standby components in the k-out-of-n system and the series system. Valdés et al. [76] studied the optimal configuration of hot standby in series system. Li and Ding [77] and Ding and Li [78] studied the hot standby unit allocation problem of k-out-of-n system.

For warm standby system, if the components are not independent and distributed, different working order of units may lead to different system reliability. Yun and Cha [79] studied the working order of units in a two-component (one primary and one standby) warm standby system with an exponential life distribution. Levitin et al. [80] studied the optimal sequence of units in 1-out-of-N warm standby system by using genetic algorithm but did not give a general conclusion. Chapter 6 of this book will focus on this issue.

References

1. Singh S, Agrafiotis G. Stochastic analysis of a two-unit cold standby system subject to maximum operation and repair time [J]. Microelectronics Reliability, 1995, 35 (12): 1489–93.
2. Kumar S, Kumar D, Mehta N. Behavioural analysis of shell gasification and carbon recovery process in a urea fertilizer plant [J]. Microelectronics Reliability, 1996, 36 (5): 671–3.
3. Pandey D, Jacob M, Yadav J. Reliability analysis of a powerloom plant with cold standby for its strategic unit [J]. Microelectronics Reliability, 1996, 36 (1): 115–9.
4. Mokaddis G, Tawfek M, Elhssia S. Analysis of a two-dissimilar unit cold standby redundant system subject to inspection and random change in units [J]. Microelectronics Reliability, 1997, 37 (2): 329–34.

5. 王冠军, 张元林. 有优先维修权和优先使用权的冷储备系统的几何过程模型 [J]. 经济数学, 2005, 22 (1): 42–9.
6. 唐应辉, 刘艳. 修理工单重休假且不能修复如新的冷储备可修系统 [J]. 数学的实践与认识, 2008, 38 (2): 47–52.
7. 余妙妙, 唐应辉, 陈胜兰. 离散时间单重休假冷储备系统的可靠性分析 [J]. 计算机工程与科学, 2008, 30 (10): 108–12.
8. 梁小林, 莫兰英, 唐小伟. 具有修理工休假的冷备退化可修系统的研究 [J]. 系统工程学报, 2010, 25 (3): 426–32.
9. Azaron A, Katagiri H, Sakawa M, et al. Reliability function of a class of time-dependent systems with standby redundancy [J]. European Journal of Operational Research, 2005, 164 (2): 378–86.
10. Azaron A, Katagiri H, Kato K, et al. Reliability evaluation of multi-component cold-standby redundant systems [J]. Applied Mathematics and Computation, 2006, 173 (1): 137–49.
11. Wang C, Xing L, Amari S V. A fast approximation method for reliability analysis of cold-standby systems [J]. Reliability Engineering & System Safety, 2012, 106 (119–26.
12. Eryilmaz S, Tank F. On reliability analysis of a two-dependent-unit series system with a standby unit [J]. Applied Mathematics and Computation, 2012, 218 (15): 7792–7.
13. Kuo W, Zuo M J. Optimal Reliability Modeling: Principles and Applications [M]. Hoboken: John Wiley & Sons, 2003.
14. Wu J S, Chen R J. An algorithm for computing the reliability of weighted-k-out-of-n systems [J]. IEEE Transactions on Reliability, 1994, 43 (2): 327–8.
15. Levitin G, Lisnianski A. Reliability optimization for weighted voting system [J]. Reliability Engineering & System Safety, 2001, 71 (2): 131–8.
16. Li W, Zuo M J. Reliability evaluation of multi-state weighted k-out-of-n systems [J]. Reliability Engineering & System Safety, 2008, 93 (1): 160–7.
17. Eryilmaz S. On reliability analysis of a k-out-of-n system with components having random weights [J]. Reliability Engineering & System Safety, 2013, 109 (0): 41–4.
18. Huang J, Zuo M J, Wu Y. Generalized multi-state k-out-of-n:G systems [J]. IEEE Transactions on Reliability, 2000, 49 (1): 105–11.
19. Tian Z, Zuo M J, Yam R C M. Multi-state k-out-of-n systems and their performance evaluation [J]. IIE Transactions, 2008, 41 (1): 32–44.
20. Coit D W, Chatwattanasiri N, Wattanapongsakorn N, et al. Dynamic k-out-of-n system reliability with component partnership [J]. Reliability Engineering & System Safety, 2015, 138: 82–92.
21. Wang K H, Kuo C C. Cost and probabilistic analysis of series systems with mixed standby components [J]. Applied Mathematical Modelling, 2000, 24 (12): 957–67.
22. Wang K H, Ke J C. Probabilistic analysis of a repairable system with warm standbys plus balking and reneging [J]. Applied Mathematical Modelling, 2003, 27 (4): 327–36.
23. Xu H, Guo W. Asymptotic stability of a parallel repairable system with warm standby [J]. International Journal of Systems Science, 2004, 35 (12): 685–92.
24. Wang K H, Lai Y J, Ke J B. Reliability and sensitivity analysis of a system with warm standbys and a repairable service station [J]. International Journal of Operations Research, 2004, 1 (1): 61–70.
25. Wang K H, Dong W L, Ke J B. Comparison of reliability and the availability between four systems with warm standby components and standby switching failures [J]. Applied Mathematics and Computation, 2006, 183 (2): 1310–22.
26. Zhang T, Xie M, Horigome M. Availability and reliability of k-out-of- (M + N):G warm standby systems [J]. Reliability Engineering & System Safety, 2006, 91 (4): 381–7.
27. Wang K H, Ke J B, Ke J C. Profit analysis of the M/M/R machine repair problem with balking, reneging, and standby switching failures [J]. Computers & Operations Research, 2007, 34 (3): 835–47.
28. Wang K H, Ke J B, Lee W C. Reliability and sensitivity analysis of a repairable system with warm standbys and R unreliable service stations [J]. The International Journal of Advanced Manufacturing Technology, 2007, 31 (11): 1223–32.

29. Shen Z, Hu X, Fan W. Exponential asymptotic property of a parallel repairable system with warm standby under common-cause failure [J]. Journal of Mathematical Analysis and Applications, 2008, 341 (1): 457–66.
30. Yuan L, Meng X Y. Reliability analysis of a warm standby repairable system with priority in use [J]. Applied Mathematical Modelling, 2011, 35 (9): 4295–303.
31. Ke J C, Wu C H. Multi-server machine repair model with standbys and synchronous multiple vacation [J]. Computers & Industrial Engineering, 2012, 62 (1): 296–305.
32. Wang K H, Yen T C, Jian J J. Reliability and sensitivity analysis of a repairable system with imperfect coverage under service pressure condition [J]. Journal of Manufacturing Systems, 2013, 32 (2): 357–63.
33. Wang K H, Liou C D, Lin Y H. Comparative analysis of the machine repair problem with imperfect coverage and service pressure condition [J]. Applied Mathematical Modelling, 2013, 37 (5): 2870–80.
34. Ke J C, Hsu Y L, Liu T H, et al. Computational analysis of machine repair problem with unreliable multi-repairmen [J]. Computers & Operations Research, 2013, 40 (3): 848–55.
35. Hsu Y L, Ke J C, Liu T H, et al. Modeling of multi-server repair problem with switching failure and reboot delay and related profit analysis [J]. Computers & Industrial Engineering, 2014, 69 (3): 21–8.
36. Mahmoud M, Esmail M. Stochastic analysis of a two-unit warm standby system with slow switch subject to hardware and human error failures [J]. Microelectronics Reliability, 1998, 38 (10): 1639–44.
37. Wang K H, Chiu L W. Cost benefit analysis of availability systems with warm standby units and imperfect coverage [J]. Applied Mathematics and Computation, 2006, 172 (2): 1239–56.
38. Hsu Y L, Ke J C, Liu T H. Standby system with general repair, reboot delay, switching failure and unreliable repair facility-A statistical standpoint [J]. Mathematics and Computers in Simulation, 2011, 81 (11): 2400–13.
39. Mokaddis G, Tawfek M. Stochastic analysis of a two-dissimilar unit warm standby redundant system with two types of repair facilities [J]. Microelectronics Reliability, 1995, 35 (12): 1467–72.
40. Pérez-Ocón R, Montoro-Cazorla D. A multiple warm standby system with operational and repair times following phase-type distributions [J]. European Journal of Operational Research, 2006, 169 (1): 178–88.
41. El-Damcese M. Analysis of warm standby systems subject to common-cause failures with time varying failure and repair rates [J]. Applied Mathematical Sciences, 2009, 3 (18): 853–60.
42. Ke J C, Wang K H. The reliability analysis of balking and reneging in a repairable system with warm standbys [J]. Quality and Reliability Engineering International, 2002, 18 (6): 467–78.
43. Jain M, Maheshwari S. N-policy for a machine repair system with spares and reneging [J]. Applied Mathematical Modelling, 2004, 28 (6): 513–31.
44. 胡兆红, 陈希镇. 两不同型部件温贮备可修系统的可靠性分析 [J]. 温州大学学报 (自然科学版), 2009, 30 (4): 44–8.
45. 刘海涛, 孟宪云, 李芳, 等. 两个不同型部件温贮备系统的几何过程模型 [J]. 系统工程, 2010, 28 (9): 103–7.
46. 杨勇, 张静文. 两个不同型多状态部件温贮备退化系统的几何过程模型 [J]. 西南师范大学学报 (自然科学版), 2013, 38 (7): 24–30.
47. 李祚东. 贮备系统的可靠性模型 [J]. 质量与可靠性, 2004, 19 (6): 29–33.
48. She J, Pecht M. Reliability of a k-out-of-n warm-standby system [J]. IEEE Transactions on Reliability, 1992, 41 (1): 72–5.
49. 李振, 张德素, 孙新利. 储备系统可靠度的归一化算法 [J]. 质量与可靠性, 2007, 22 (6): 13–5.
50. Amari S V, Pham H, Misra R B. Reliability characteristics of k-out-of-n warm standby systems [J]. IEEE Transactions on Reliability, 2012, 61 (4): 1007–18.
51. Amari S V, Dill G. A new method for reliability analysis of standby systems [A]. Annual Reliability and Maintainability Symposium (RAMS2009), Fort Worth, Texas, USA, 2009 [C]. Piscataway, NJ: IEEE, 417–22.

52. Tannous O, Xing L, Dugan J. Reliability analysis of warm standby systems using sequential BDD [A]. Annual Reliability and Maintainability Symposium (RAMS2011), Lake Buena Vista, Florida, USA,2011 [C]. Piscataway, NJ: IEEE, 1–7.
53. Tannous O, Xing L. Efficient analysis of warm standby systems using central limit theorem [A]. Annual Reliability and Maintainability Symposium (RAMS2012),2012 [C]. Piscataway, NJ: IEEE, 1–6.
54. Levitin G, Amari S V. Approximation algorithm for evaluating time-to-failure distribution of k-out-of-n system with shared standby elements [J]. Reliability Engineering & System Safety, 2010, 95 (4): 396–401.
55. Cha J H, Mi J, Yun W Mo Ydelling a general standby system and evaluation of its performance [J]. Applied Stochastic Models in Business and Industry, 2008, 24 (2): 159–69.
56. Li X, Zhang Z, Wu Y. Some new results involving general standby systems [J]. Applied Stochastic Models in Business and Industry, 2009, 25 (5): 632–42.
57. Papageorgiou E, Kokolakis G. Reliability analysis of a two-unit general parallel system with (n-2) warm standbys [J]. European Journal of Operational Research, 2010, 201 (3): 821–7.
58. Eryilmaz S. Reliability of a k-out-of-n system equipped with a single warm standby component [J]. IEEE Transactions on Reliability, 2013, 62 (2): 499–503.
59. Kuo W, Prasad V R. An annotated overview of system-reliability optimization [J]. IEEE Transactions on Reliability, 2000, 49 (2): 176–87.
60. Coit D W. Cold-standby redundancy optimization for nonrepairable systems [J]. IIE Transactions, 2001, 33 (6): 471–8.
61. Coit D W. Maximization of system reliability with a choice of redundancy strategies [J]. IIE Transactions, 2003, 35 (6): 535–43.
62. Coit D W, Jin T, Wattanapongsakorn N. System optimization with component reliability estimation uncertainty: a multi-criteria approach [J]. IEEE Transactions on Reliability, 2004, 53 (3): 369–80.
63. Boddu P, Xing L. Redundancy allocation for k-out-of-n: G systems with mixed spare types [A]. Annual Reliability and Maintainability Symposium (RAMS2012), 2012 [C]. IEEE, 1–6.
64. Boddu P, Xing L. Reliability evaluation and optimization of series-parallel systems with k-out-of-n: G subsystems and mixed redundancy types [J]. Proceedings of the Institution of Mechanical Engineers, Part O: Journal of Risk and Reliability, 2013, 227 (2): 187–98.
65. Yu H, Yalaoui F, Châtelet É, et al. Optimal design of a maintainable cold-standby system [J]. Reliability Engineering & System Safety, 2007, 92 (1): 85–91.
66. Huang H Z, Qu J, Zuo M J. Genetic-algorithm-based optimal apportionment of reliability and redundancy under multiple objectives [J]. IIE Transactions, 2009, 41 (4): 287–98.
67. Azaron A, Perkgoz C, Katagiri H, et al. Multi-objective reliability optimization for dissimilar-unit cold-standby systems using a genetic algorithm [J]. Computers & Operations Research, 2009, 36 (5): 1562–71.
68. Safari J. Multi-objective reliability optimization of series-parallel systems with a choice of redundancy strategies [J]. Reliability Engineering & System Safety, 2012, 108 (10–20.
69. Abouei Ardakan M, Zeinal Hamadani A, Alinaghian M. Optimizing bi-objective redundancy allocation problem with a mixed redundancy strategy [J]. ISA Transactions, 2015, 55 (0): 116–28.
70. Amari S V, Dill G. Redundancy optimization problem with warm-standby redundancy [A]. Annual Reliability and Maintainability Symposium (RAMS2010), 2010 [C]. IEEE, 2010: 1–6.
71. Tannous O, Xing L, Peng R, et al. Redundancy allocation for series-parallel warm-standby systems [A]. Proceedings of IEEE International Conference on Industrial Engineering and Engineering Management (IEEM), Singapore, 2011 [C]. IEEE, 2011: 1261–5.
72. Boland P J, El-Neweihi E, Proschan F. Stochastic order for redundancy allocations in series and parallel systems [J]. Advances in Applied Probability, 1992, 24 (1): 161–71.
73. Romera R, Valdés J E, Zequeira R I. Active-redundancy allocation in systems [J]. IEEE Transactions on Reliability, 2004, 53 (3): 313–8.
74. Li X, Hu X. Some new stochastic comparisons for redundancy allocations in series and parallel systems [J]. Statistics & Probability Letters, 2008, 78 (18): 3388–94.

75. Hu T, Wang Y. Optimal allocation of active redundancies in r-out-of-n systems [J]. Journal of Statistical Planning and Inference, 2009, 139 (10): 3733–7.
76. Valdés J E, Arango G, Zequeira R I, et al. Some stochastic comparisons in series systems with active redundancy [J]. Statistics & Probability Letters, 2010, 80 (11): 945–9.
77. Li X, Ding W. Optimal allocation of active redundancies to k-out-of-n systems with heterogeneous components [J]. Journal of Applied Probability, 2010, 47 (1): 254–63.
78. Ding W, Li X. The optimal allocation of active redundancies to k-out-of-n systems with respect to hazard rate ordering [J]. Journal of Statistical Planning and Inference, 2012, 142 (7): 1878–87.
79. Yun W Y, Cha J H. Optimal design of a general warm standby system [J]. Reliability Engineering & System Safety, 2010, 95 (8): 880–6.
80. Levitin G, Xing L, Dai Y. Optimal sequencing of warm standby elements [J]. Computers & Industrial Engineering, 2013, 65 (4): 570–6.

Chapter 2
Related Concepts in Reliability Modeling of Warm Standby Systems

From the reliability modeling perspective, the warm standby component can fail both in the warm standby state and the normal working state, and its lifetime in the normal working state may depend on the duration of the warm standby state. Consequently, the whole lifetime of warm standby component is dependent on the switch time, which is related to the failures of the online components. Therefore, the lifetime of the components in the warm standby system cannot be treated as independent, and the reliability modeling of warm standby system is not a simple enumeration of failure combinations but must take the component switching into account.

In the warm standby system, whenever a component failure occurs, the system has to detect and isolate the faulty component, after which the standby components can be switched to the online mode. The fault detection and isolation cannot be perfect as there is an opportunity that the system may miss or misspecify the fault [1–5]. Therefore, the imperfect fault coverage effect should be considered in the reliability modeling of warm standby systems. In addition, the switching process may also fail, for example, due to the failure of the switch. As a summary, the operation of warm standby systems is a complex dynamic process, and its reliability modeling is more challenging than the cold or hot standby systems.

2.1 Reliability Models for Warm Standby Component

A warm standby component may experience three possible states, i.e., warm standby, normal working and failed, as shown in Fig. 2.1. At the beginning, the warm standby state stays in the warm standby state and may be switched to the normal working state or fail before switching at some time point. If the warm standby component is switched to the online working state, it may also fail at some future time. Note that the switching of the warm standby component is dependent on the other components in the system. Therefore, it needs to consider the failure of other components, rather than

R. Peng et al., *Reliability Modelling and Optimization of Warm Standby Systems*,
https://doi.org/10.1007/978-981-16-1792-8_2

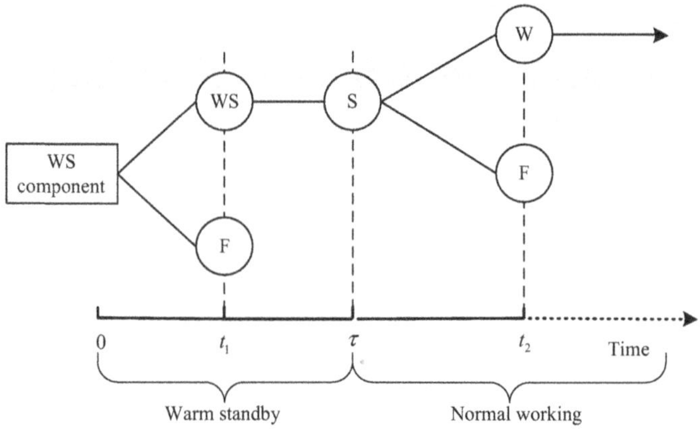

Fig. A possible life of a warm standby component (WS: warm standby; S: switching; W: normal working; F: failed)

view the lifetime of the warm standby component is independent to other components in the system when doing reliability modeling.

Because the working loads in the warm standby state and the normal working state are different, the failure patterns and the lifetime distributions for the warm standby component may be different in these two states. This book considers two possible assumptions on the lifetime distributions of a warm standby component in the two different states:

(1) The failure mechanism in the warm standby state is the same as that in the normal working state. In this case, it is reasonable to assume that the lifetime distribution in the warm standby state has the same form as that in the normal working state, except that the aging in the warm standby state is slower or the lifetime in the normal working state is accelerated. Some commonly used models for accelerated lifetime include the accelerated failure time model (AFTM) [6–8] and the proportional hazard model (PHM) [9–11].

Denote the lifetime distribution in the normal working state by $F^{\mathrm{o}}(\cdot)$, and the lifetime distribution in the warm standby state by $F^{\mathrm{s}}(t)$, where the superscript "o" stands for the operational state and "s" stands for the warm standby state. If we exploit the linear AFTM, and assume that the lifetime in the warm standby state is shrunk through a constant γ compared with that in the normal working state, then we can model the lifetime distribution for this component in the warm standby state as $F^{\mathrm{s}}(t) = F^{o}(\gamma t)$. Here, it is generally required that $0 < \gamma < 1$. The interpretation for this model is that the aging of the component at t in the warm standby state is identical to the aging of the same component at γt in the normal working state. This model can be generalized to include nonlinear acceleration effect:

$$F^{\mathrm{s}}(t) = F^{\mathrm{o}}(\gamma(t)), \tag{2.1}$$

where $\gamma(t) \leq t$ is a monotonically non-decreasing function with $\gamma(0) = 0$ and $\lim_{t\to+\infty} \gamma(t) = +\infty$. More complicated relationship between the lifetimes in the warm standby state and the normal working state can be modeled using such nonlinear $\gamma(t)$.

Since the warm standby component may experience the warm standby and the normal working states during its lifetime, we should also model the dependence of the lifetimes in different states. In step-stress acceleration test, three commonly used models to imbed the load history are the tampered random variable (TRV) [12], the cumulative exposure model (CEM) [13, 14] and the tampered failure rate (TFR) [15, 16]. The CEM is also called the cumulative damage model, which suggests to transform the durations in different loads into a comparable effective age in some baseline load. In this way, the effective age of a component at any point of time is equivalent to the sum of the historical load durations times the corresponding acceleration factors. The effective age is addictive, which is irrelevant to the sequence of loads. The remaining life of the component is only dependent on the current stress and the current cumulated damage (effective age), regardless of the damage accumulation history. Meanwhile, the change of load level will only change the distribution of the remaining life, but not the cumulated damage.

For the warm standby system, we can use a general AFTM to link the lifetimes between the warm standby state and the normal working state and apply a CEM to model the whole life of the warm standby component considering state switching. Specifically, we assume that the cumulative damage at τ under the warm standby state is $F^s(\tau) = F^o(\gamma(\tau))$, i.e., the effective age of the component at τ is $\gamma(\tau)$ if the normal working state is viewed as the baseline state. If the component is switched to the normal working state at time τ, then the remaining life would follow the distribution $F^o(\cdot)$ with an initial damage $F^o(\gamma(\tau))$. This is equivalent to say that the failure probability in the normal working state after switching at τ is the same as if the same component has been operated a duration of $\gamma(\tau)$ under the normal working state. Therefore, the lifetime distribution for the warm standby component considering switching is:

$$F(t) = \begin{cases} F^o(\gamma(t)), 0 < t \leq \tau \\ F^o(t - \tau + \gamma(\tau)), t > \tau \end{cases} \tag{2.2}$$

Thus, with a given $\gamma(\cdot)$, the lifetime distribution can be described by the cumulative distribution function (CDF) under the normal working condition $F^o(t)$ conditional on the switching time τ. Moreover, conditional on that the component is switched to the online state at τ, the remaining lifetime at the normal working state is as follows:

$$F_\tau^o(u) = \frac{F^o(u + \gamma(t))}{1 - F^o(\gamma(t))}, u > 0,$$

where u is the lifetime of the component in the normal working state.

By using other models, e.g., the TRV model, similar models for the warm standby component can be obtained. Nevertheless, we would only consider the AFTM and CEM in the book and leave the other models to the readers.

(2) The failure mechanism in the warm standby state is different from that in the normal working state, and they are independent. In this case, the lifetime in the warm standby state is independent from that in the normal working state, and we can assume two independent lifetime distributions $F^{o}(t)$ and $F^{s}(t)$ for the normal working state and the warm standby state, respectively. Conditional on the switching time τ, the failure probability at any time t is

$$F(t) = \begin{cases} F^{s}(t), 0 < t \leq \tau \\ F^{o}(t-\tau)[1 - F^{s}(\tau)], t > \tau \end{cases}. \tag{2.3}$$

2.2 Common Structures for Warm Standby Systems

Warm standby belongs to a kind of redundancy design, which often has a k-out-of-n voting mechanism. As shown in Fig. 2.2, such system generally consists of n components with same function, and the system can satisfy the demand as long as there are $k(k \leq n)$ working in the normal state. Whenever any one of the components fails, the system would switch a standby component on to replace the failed component so that the system always has k components in the normal working state.

More generally, each component in the system can have a nominal capacity, while the system operation requires that the total capacity of working components must meet a certain criterion, instead of requiring that the number of working components meet a threshold. For example, the power of a local electrical grid depends on the total capacity of the power plants instead on the number of the power plants. Such a system can be viewed as a generalization of the k-out-of-n system, in that it would

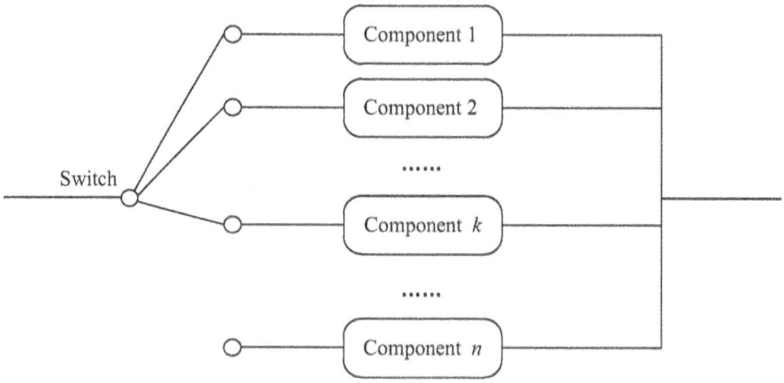

Fig. 2.2 A commonly met k-out-of-n redundant design

degenerate to a k-out-of-n system if all the components have the same capacity. For a general system that the components have different capacities, it is different from the k-out-of-n system because there is no such "k" as the threshold for the number of working components. Since the system operation characteristics is dependent on the specific demand for the system, such systems can be named as the demand-based systems.

The k-out-of-n systems and the demand-based systems are general structures for warm standby systems, and we would mainly consider these systems in the following parts of the book.

2.3 Imperfect Fault Coverage

Whenever a component failure occurs, the system has to detect and isolate the failed component. Namely, the system has to cover the fault. The so-called fault coverage is that the system can still function properly after some faults occur in the system [3]. A typical diagram for fault coverage is illustrated in Fig. 2.3.3 [17]. When a component fail occurs, there are three possible outcomes:

- If the fault is transient, and it can recover to the normal state without isolate any component, then this fault has an outcome "Restoration (R)";
- If the fault is permanent and the faulty component is correctly isolated, then this fault corresponds to the outcome "Fault-coverage (C)";
- If the component failure leads to a system break-down, then this scenario has an outcome "Single-point failure (S)"[18].

In system reliability analysis, assume the three possible scenarios have probabilities r_0, s_0 and c_0, corresponding to R, S and C, respectively. Then we have

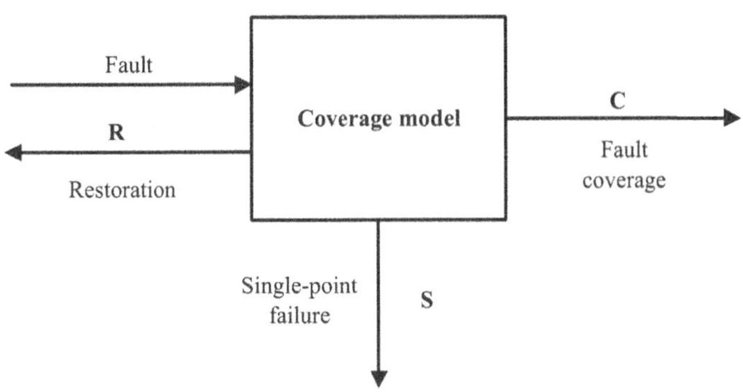

Fig. 2.3 A general model for fault coverage

$$r_0 + s_0 + c_0 = 1.$$

In general, a transient failure with "Restoration" as the outcome can be excluded as a failure, and we can assume $r_0 = 0$. Then, if a component failure occurs, it would be covered or lead to a system failure. The fault coverage probability is the probability that a fault is properly covered given it occurred:

$$c_0 = \Pr\{Systemnormal(faultcovered)|componentfailureoccurs\} \tag{2.4}$$

The fault coverage concept was first introduced by Bouricius et al. (1969) [3]. Arnold (1973) [4] introduced the imperfect fault coverage into the availability modeling of repairable systems. The studies on imperfect fault coverage include: (1) modeling the fault coverage mechanism and procedure; and (2) incorporating this concept in the system reliability modeling. In the system reliability modeling, some early studies viewed the fault coverage probability as a constant and the fault tree and the reliability block diagram can be used to model the system reliability considering imperfect fault coverage. For example, Dugan (1989) [19] studied how to model the imperfect fault coverage in reliability modeling using fault tree analysis. In reality, the fault detection, isolation and system restoration are random events, and they depend on the system state. Therefore, the fault coverage is in fact a dynamic, sequential process. Some studies use Markov chain to model such fault coverage process. Nevertheless, the shortcoming of the Markov chain method is that the state space would suffer the state space explosion problem, which inhibits its application to large-scale systems. Some studies combined the advantages of combinatory methods and the Markov chain, and proposed hybrid approaches such as hybrid automated reliability predictor (HARP). When using combinatory methods to incorporate the fault coverage effect, the key is to "block" the imperfect fault coverage effect from the system modeling. In this context, Amari et al. (1999) [17] elegantly exploited such a methodology and proposed the simple efficient analysis (SEA) method. The key observation of the SEA method is that there are three possible states for a component if the imperfect fault coverage is considered, i.e., normal, failed and covered, and failed but not covered. When assuming that every uncovered fault would lead to system failure, the system reliability is equal to the probability that all the faults are covered and the system's function satisfies the design requirement. This probability can be decomposed as.

$$R_S = \Pr\{System\,reliable|All\,the\,faults\,are\,covered\}\cdot\Pr\{All\,the\,faults\,are\,covered\}.$$

Here, the second term on the right-hand side, i.e., the probability that all the faults are covered can be calculated from a simple combination of the fault coverage probabilities of all the components. The first term is obtained in two steps: (1) calculating the reliability or failure probability of each component conditional on the fault can be covered; and (2) calculating the system reliability or the system failure probability based on these quantities for all the components. In the second step, any existing methods for system reliability assessment can be applied. As a conclusion, the key point in SEA is to isolate the imperfect fault coverage effect so that the system reliability can be calculated without considering the imperfect fault coverage effect.

Later, Amari et al. (2007) [20] further proposed two approaches to incorporate the imperfect fault coverage into the system reliability analysis based on SEA. Chen et al. (2005) [21] extended the SEA method into the reliability analysis of hierarchical systems and proposed a top-down separable algorithm for such systems. Chen et al. (2006, 2007) [22, 23] proposed to combine the SEA method with Markov chain to address the reliability modeling problem for phased-mission systems with imperfect fault coverage.

In the SEA method, each component has a fixed coverage probability, which is called the element level coverage (ELC) model. The coverage probability for each component is predetermined in the ELC model, and different components may have different coverage probabilities. This model is suitable for the fault coverage modeling of the components with built-in test. Another fault coverage model is the fault level coverage (FLC) model, where the coverage probability of a fault only depends on the order of the fault in the fault sequence instead of the failed component itself. In this model, the first fault in the system is covered with probability c_1, the second fault is covered with probability c_2 and so on. This model is suitable for the coverage modeling of redundant systems with middle value voting mechanisms [24]. For redundant systems with voting mechanisms, the faulty component can be automatically detected by comparing its state with others, and the coverage probability is relevant to the number of existing normal components. Levitin and Amari (2008) [25] studied the system reliability modeling of series–parallel systems considering the FLC effect, and proposed an approach based on reliability block diagram and universal generating functions for system reliability modeling. Myers (2007) [24] studied the reliability of k-out-of-n systems considering FLC, and Levitin and Amari (2009) [26] proposed methods to evaluate the system reliability of series–parallel systems based on universal generating functions. A review on the imperfect fault coverage models and the system reliability analysis considering imperfect fault coverage can be found in [27].

2.4 Decision Diagram

2.4.1 Binary Decision Diagram

Binary decision diagram (BDD) is a directed acyclic graph based on Shannon decomposition [28]. Shannon decomposition is a transformation of Boolean functions:

$$f = x \cdot f_{x=1} + \overline{x} \cdot f_{x=0} = ite(x, f_{x=1}, f_{x=0}), \tag{2.5}$$

where f is a Boolean function of a set of variables $\mathbf{X}$ and x is an element of $\mathbf{X}$. i te stands for the structure "If…Then…Else…". The two terms $x \cdot f_{x=1}$ and $\overline{x} \cdot f_{x=0}$ correspond to the two possible scenarios when x is true and false, respectively. The two events are mutually exclusive.

To illustrate the Shannon decomposition in reliability modeling, let us consider a two-component parallel system. Denote the states of the two components by x_1 and x_2, with "1" standing for failure and "0" standing for normal. Let f denotes the system state. The Boolean function for the system is

$$f = x_1 x_2., \tag{2.6}$$

which takes value 1 if and only if $x_1 = x_2 = 1$. The Shannon decomposition can decompose this function f into two subfunctions:

$$\begin{aligned} f &= x_1 f_{x_1=1} + \overline{x}_1 f_{x_1=0}, \\ f_{x_1=1} &= x_2,\ f_{x_1=0} = 0 \end{aligned} \tag{2.7}$$

With Shannon decomposition, a function can be expanded or developed to any desired depth so that the analysis of Boolean systems can be greatly simplified.

The fault tree analysis is a classic and useful tool for reliability modeling and many software packages contain this tool. This method provides a top-down, deductive approach to analyze the system functionality. However, it is not straightforward to evaluate the system reliability using the fault tree, as it needs the enumeration of the cut sets which is often a cumbersome task. This is especially challenging for large-scale systems. This problem can be addressed by transforming the fault tree to a BDD structure, which is easier and more convenient for computer to handle. A fault tree with OR and AND gates can be transformed to a BDD in a bottom-up way as follows [28]:

$$\begin{aligned} G\Diamond H &= \text{ite}(x, G_1, G_2)\Diamond\text{ite}(y, H_1, H_2) \\ &= \begin{cases} ite(x, G_1\Diamond H_1, G_2\Diamond H_2) index(x) = index(y) \\ ite(x, G_1\Diamond H, G_2\Diamond H) index(x) < index(y) \\ ite(y, G\Diamond H_1, G\Diamond H_2) index(x) > index(y) \end{cases} \end{aligned} \tag{2.8}$$

where G and H represent two subtrees, the symbol "$\diamond$" represents the logic operator "AND" or "OR", and index(·) represents the ordering of a Boolean variable in the variable table.

Consider a series–parallel system with tree components, as shown in Fig. 2.4. The system has two subsystems, and the system fails if any one of the subsystems fails. If we denote the system failure by F, and the left and right trees of the fault tree by G_1 and G_2, respectively, then we will have $F = G_1 \text{OR} G_2$. Further, we can note that the left subtree G_1 can be decomposed as $G_1 = H_1 \text{AND} H_2$, where $H_1 = \text{ite}(x_1, 1, 0)$ corresponding to the state of component A_1 and $H_2 = \text{ite}(x_2, 1, 0)$ corresponding

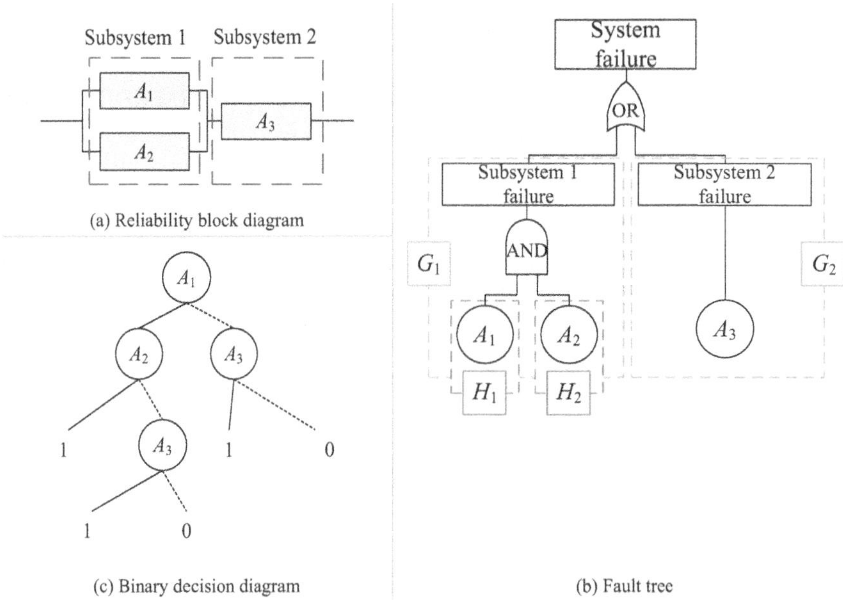

Fig. 2.4 The reliability block diagram, fault tree and binary decision diagram for a series–parallel system

to the state of component A_2. Similarly, the right subtree G_2 can be represented by $G_2 = \text{ite}(x_3, 1, 0)$, meaning that G_2 equals 1 if component A_3 fails. Define the ordering of the three Boolean variables as $x_1 < x_2 < x_3$ in the variable table, then from (1–8) we shall have

$$
\begin{aligned}
G_1 =& H_1 AND H_2 = ite(x_1, 1\ AND H_2, 0\ AND H_2) = ite(x_1, H_2,\ 0) = ite(x_1, ite(x_2, 1, 0), 0),\\
F =& G_1 OR G_2 = ite(x_1, H_2 OR G_2, 0\ OR G_2) = ite(x_1, ite(x_2, 1\ OR G_2, 0\ OR G_2), G_2)\\
=& ite(x_1, ite(x_2, 1, G_2), G_2) = ite(x_1, ite(x_2, 1, ite(x_3, 1, 0))\ , ite(x_3, 1, 0))
\end{aligned}
$$

In such a manner, we obtain the Shannon decomposition for the top event F as a Boolean function. Corresponding to this decomposition, the BDD representation is obtained straightforwardly, as shown in Fig. 2.4c.

Based on the transformation rules of the BDD, the terminal node represents the system state. In particular, if "1" and "0" represent the system failure and normal states, respectively, then every path with terminal node equaling to "1" corresponds to a possible combination of component states that leads to system failure. Based on the rules of Shannon decomposition and the structure of the BDD, the state combinations indicated by different paths are exclusive and exhaustive. Therefore, the system failure probability can be obtained as the sum of the occurrence probabilities of all the paths that lead to the terminal "1".

2.4.2 Multi-Valued Decision Diagram

The multi-valued decision diagram (MDD) or multi-state MDD (MMDD) method is an extension of the BDD method, which was originally proposed to analyze static multi-state fault trees [29–32]. It has two sink nodes, labeled "0" and "1", respectively representing a binary-state system being in the operational and failed states, or a multi-state system being in a certain state or not. A non-sink node in the MDD has multiple outgoing edges, each corresponding to a state of the multi-state component represented by the node. For example, an aircraft engine can fail in different stage of its mission, including taxi, ascent, level-flight, descent and landing phases, and may also complete the mission without fail [33]. Then, we can use an MDD with six branches to represent the state of the engine, as shown in Fig. 2.5 (a), where "1" stands for failure and "0" stands for normal.

In general, a non-sink node representing a multi-state component A with k states is associated with an k-valued state variable x_A, and has k outgoing edges $\{1, 2, \ldots, k\}$. The logic expression for x_A can be represented in MDD using the *case* format as

$$\begin{aligned} F &= x_A|_1 \cdot F_{x_A=1} + x_A|_2 \cdot F_{x_A=2} + \cdots + x_A|_k \cdot F_{x_A=k} \\ &= case(x_A, F_1, F_2, \ldots, F_k), \end{aligned} \tag{2.9}$$

where $F_i \triangleq F_{x_A=i}$. Similar to the if–then-else (*ite*) format of the Boolean expression in the binary decision diagram operation, based on the Shannon's decomposing rule, a multi-valued logic expression F can be decomposed into k sub-expressions with regard to any of its constituent variable, e.g., x_A as shown in (1–9). Each value i of x_A corresponds to a sub-expression $F_{x_A=i}$ which is F evaluated at x_A being i. All the sub-expressions $x_A|_i \cdot F_{x_A=i}$ in (1) are disjoint.

The MDD for a multi-state fault trees can be generated recursively by applying manipulation rules of (1–10):

$$G \Diamond H = case(x, G_1, \ldots, G_m) \Diamond case(y, H_1, \ldots, H_m)$$

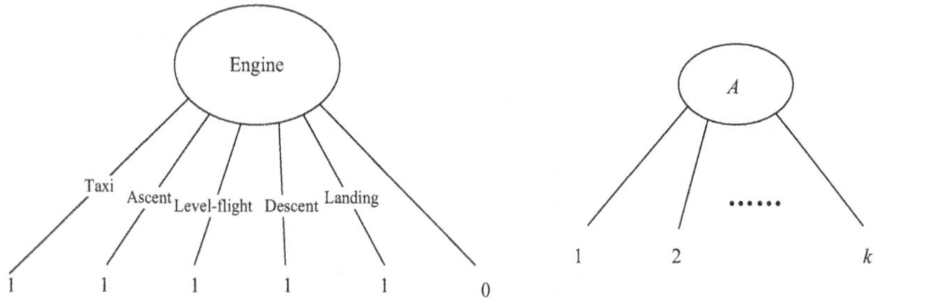

(a) MDD representation for an engine

(b) A general MDD for a k-state component

Fig. 2.5 MDD for the engine and general cases

$$= \begin{cases} case(x, G_1 \diamondsuit H_1, \ldots, G_m \diamondsuit H_m) index(x) = index(y) \\ case(x, G_1 \diamondsuit H, \ldots, G_m \diamondsuit H) index(x) < index(y) \\ case(y, G \diamondsuit H_1, \ldots, G \diamondsuit H_m) index(x) > index(y) \end{cases}, \quad (2.10)$$

where G and H represent two multi-valued expressions corresponding to the traversed sub-fault trees. The logical operation (AND, OR) is represented by $\diamond$.

The probability of the system being in a particular state can be calculated as the sum of probabilities of all the disjoint paths from the root node to the sink node "1". These paths represent all possible combinations of component states that lead to the system being in a particular state.

2.4.3 Decision Diagram in Reliability Modeling

As an effective modeling approach, the decision diagram has been widely used in system reliability modeling. The BDD was first proposed by Lee (1959) [34] and developed further by Boute (1976) [35] and Akers (1978) [36]. Bryant (1986) [28] gave an in-depth discussion to the existing binary decision diagram and proposed the ordered BDD (OBDD) and the rules for simplification. Since then, the BDD is widely applied in circuit verification [37], compact Markov chain representation [38–40], reachability set generation and storage for Petri nets [41, 42] and other related fields.

The BDD was used in system reliability analysis since the beginning of 1990s [30, 43]. Zang et al. (2003) [44] proposed a reliability evaluation approach for multi-state systems based on BDD method. Chang et al. (2005) [18] proposed to apply OBDD to analyze the reliability of multi-state systems considering imperfect fault coverage. Rauzy et al. (2007) [45] studied the method to transform large-scale fault trees to BDD. Myers and Rauzy (2008) [46, 47] proposed to use BDD to evaluate the reliability of k-out-of-n systems under imperfect fault coverage. Shrestha and Xing (2008) [48] proposed a logarithmic BDD-based method for reliability analysis of multi-state systems. Rauzy (2008) [49] gave an overview of the BDD in reliability analysis in 15 years and summarized the BDD fault tree analysis and related probability calculation. Min et al. (2005) [50], Zhang et al. (2007) [51], Tao et al. (2009) [52], Hu (2010) [53], and Gao and Zhang (2011) [54] discussed using BDD to analyze fault trees. To overcome the shortcomings in analyzing fault trees using BDD, Tu (2010) [55] proposed to decompose the fault tree into subtrees and analyze each subtree using BDD. As a generalization of BDD, the MDD was first used to analyze the multi-state system reliability [29–32]. Some later studies extended the MDD to the reliability modeling of the phased-missing systems [56, 57]. Mo and Xing (2013) [58] used BDD and MDD to analyze the system reliability considering common-cause failure (CCF). Xing and Amari gave a comprehensive discussion on the applications of decision diagrams in system reliability modeling in their book *Binary Decision Diagrams and Extensions for System Reliability Analysis* [59], and we refer the interested readers to this book.

References

1. Dobson I, Carreras B A, Lynch V E, et al. Complex systems analysis of series of blackouts: Cascading failure, critical points, and self-organization [J]. Chaos, 2007, 17 (2): 026103 (1–13).
2. Pepyne D L. Topology and cascading line outages in power grids [J]. Journal of Systems Science and Systems Engineering, 2007, 16 (2): 202–21.
3. Bouricius W, Carter W C, Schneider P. Reliability modeling techniques for self-repairing computer systems [A]. The 24th National ACM Conference, 1969 [C]. ACM, 1969: 295–309.
4. Arnold T F. The concept of coverage and its effect on the reliability model of a repairable system [J]. IEEE Transactions on Computers, 1973, 100 (3): 251–4
5. Xing L. Reliability evaluation of phased-mission systems with imperfect fault coverage and common-cause failures [J]. IEEE Transactions on Reliability, 2007, 56 (1): 58–68.
6. Kay R, Kinnersley N. On the use of the accelerated failure time model as an alternative to the proportional hazards model in the treatment of time to event data: A case study in influenza [J]. Drug Information Journal, 2002, 36 (3): 571–9.
7. Finkelstein M. On statistical and information-based virtual age of degrading systems [J]. Reliability Engineering & System Safety, 2007, 92 (5): 676–81.
8. Barabadi A, Barabady J, Markeset T. Application of accelerated failure model for the oil and gas industry in Arctic region [A]. Proceedings of IEEE International Conference on Industrial Engineering and Engineering Management (IEEM), 2010 [C].IEEE, 2244–8.
9. Cox D R. Regression models and life-tables [J]. Journal of the Royal Statistical Society Series B (Methodological), 1972, 34 (2): 187–220.
10. Kumar D, Klefsjö B. Proportional hazards model: a review [J]. Reliability Engineering & System Safety, 1994, 44 (2): 177–88.
11. Li X, Yan R, Zuo M J. Evaluating a warm standby system with components having proportional hazard rates [J]. Operations Research Letters, 2009, 37 (1): 56–60.
12. Nelson W. Accelerated life testing step-stress models and data analysis [J]. IEEE Transaction on Reliability, 1980, 29: 103–108.
13. DeGroot M, Goel P. Bayesian estimation and optimal designs in partially accelerated life testing [J]. Naval Research Logistics Quarterly, 1979, 26: 223–235.
14. Amari S V, Bergman R. Reliability analysis of k-out-of-n load-sharing systems [A]. Annual Reliability & Maintainability Symposium (RAMS2008), LA, USA, 2008 [C]. Piscataway, NJ: IEEE, 440–445.
15. Solomon P. Effect of misspecification of regression models in the analysis of survival data [J]. Baometrzka, 1984, 71 (2): 291–298.
16. Nelson W. Accelerated Testing: Statistical Models, Test Plans, and Data Analyses. Wiley & Sons, 1990.
17. Amari S V, Dugan J B, Misra R B. A separable method for incorporating imperfect fault-coverage into combinatorial models [J]. IEEE Transactions on Reliability, 1999, 48 (3): 267–74.
18. Chang Y R, Amari S V, Kuo S Y. OBDD-based evaluation of reliability and importance measures for multistate systems subject to imperfect fault coverage [J]. IEEE Transactions on Dependable and Secure Computing, 2005, 2 (4): 336–47.
19. Dugan J B. Fault trees and imperfect coverage [J]. IEEE Transactions on Reliability, 1989, 38 (2): 177–85.
20. Amari S, Myers A, Rauzy A. An efficient algorithm to analyze new imperfect fault coverage models [A]. Annual Reliability and Maintainability Symposium (RAMS2007), 2007 [C].IEEE, 420–6.
21. 陈光宇, 黄锡滋, 唐小我. 不完全覆盖的多层次系统可靠性分析 [J]. 系统工程学报, 2005, 20 (5): 60–6.
22. 陈光宇, 黄锡滋, 唐小我. 不完全覆盖的多阶段任务系统可靠性集成分析 [J]. 系统工程理论与实践, 2006, 26 (4): 1–8.
23. 陈光宇, 黄锡滋, 张小民, et al. 不完全覆盖的多阶段任务系统可靠性综合分析 [J]. 系统工程学报, 2007, 22 (5): 539–45.

24. Myers A F. k-out-of-n: G system reliability with imperfect fault coverage [J]. IEEE Transactions on Reliability, 2007, 56 (3): 464–73.
25. Levitin G, Amari S V. Multi-state systems with multi-fault coverage [J]. Reliability Engineering & System Safety, 2008, 93 (11): 1730–9.
26. Levitin G, Amari S. Three types of fault coverage in multi-state systems [A]. 8th International Conference on Reliability, Maintainability and Safety (ICRMS 2009), 2009 [C]. IEEE, 122–7.
27. Amari S V, Myers A F, Rauzy A, et al. Imperfect coverage models: status and trends [M]//MISRA K B. Handbook of Performability Engineering. London; Springer. 2008: 321–48.
28. Bryant R E. Graph-based algorithms for Boolean function manipulation [J]. IEEE Transactions on Computers, 1986, 100 (8): 677–91.
29. Akers J, Bergman R, Amari S V, et al. Analysis of multi-state systems using multi-valued decision diagrams [A]. Annual Reliability and Maintainability Symposium (RAMS2008), 2008 [C]. Piscataway, NJ: IEEE, 347–53.
30. Xing L, Dai Y. A new decision-diagram-based method for efficient analysis on multistate systems [J]. IEEE Transactions on Dependable and Secure Computing, 2009, 6 (3): 161–74.
31. Shrestha A, Xing L, Coit D W. An efficient multistate multivalued decision diagram-based approach for multistate system sensitivity analysis [J]. IEEE Transactions on Reliability, 2010, 59 (3): 581–92.
32. Amari S V, Xing L, Shrestha A, et al. Performability analysis of multistate computing systems using multivalued decision diagrams [J]. IEEE Transactions on Computers, 2010, 59 (10): 1419–33.
33. Xing L, Amari S V. Reliability of phased-mission systems [M]//MISRA K B. Handbook of Performability Engineering. Springer. 2008: 349–68.
34. Lee C-Y. Representation of switching circuits by binary-decision programs [J]. Bell System Technical Journal, 1959, 38 (4): 985–99.
35. Boute R T. The binary decision machine as programmable controller [J]. Euromicro Newsletter, 1976, 2 (1): 16–22.
36. Akers S B. Binary decision diagrams [J]. IEEE Transactions on Computers, 1978, 100 (6): 509–16
37. Burch J R, Clarke E M, Long D E, et al. Symbolic model checking for sequential circuit verification [J]. IEEE Transactions on Computer-Aided Design of Integrated Circuits and Systems, 1994, 13 (4): 401–24.
38. Hermanns H, Meyer-Kayser J, Siegle M. Multi terminal binary decision diagrams to represent and analyse continuous time Markov chains [A]. 3rd Int Workshop on the Numerical Solution of Markov Chains, 1999 [C]. Citeseer, 188–207.
39. Ciardo G, Lüttgen G, Siminiceanu R. Saturation: An efficient iteration strategy for symbolic state-space generation [A]. Proceedings of the 7th International Conference on Tools and Algorithms for the Construction and Analysis of Systems, 2001 [C].Springer-Verlag, 328–42.
40. Miner A S, Cheng S. Improving efficiency of implicit Markov chain state classification [A]. Proceedings of First International Conference on the Quantitative Evaluation of Systems (QEST 2004), 2004 [C]. IEEE, 262–71.
41. Miner A S, Ciardo G. Efficient reachability set generation and storage using decision diagrams [A]. International Conference on Application and Theory of Petri Nets, 1999 [C]. Springer-Verlag, 6–25.
42. Ciardo G. Reachability set generation for Petri nets: Can brute force be smart? [A]. 25th International Conference on Application and Theory of Petri Nets, Bologna, Italy,2004 [C]. Springer, 17–34.
43. Rauzy A. New algorithms for fault trees analysis [J]. Reliability Engineering & System Safety, 1993, 40 (3): 203–11.
44. Zang X, Wang D, Sun H, et al. A BDD-based algorithm for analysis of multistate systems with multistate components [J].IEEE Transactions on Computers, 2003, 52 (12): 1608–18.
45. Rauzy A, Gauthier J, Leduc X. Assessment of large automatically generated fault trees by means of binary decision diagrams [J]. Proceedings of the Institution of Mechanical Engineers, Part O: Journal of Risk and Reliability, 2007, 221 (2): 95–105

46. Myers A F, Rauzy A. Assessment of redundant systems with imperfect coverage by means of binary decision diagrams [J]. Reliability Engineering & System Safety, 2008, 93 (7): 1025–35
47. Myers A, Rauzy A. Efficient reliability assessment of redundant systems subject to imperfect fault coverage using binary decision diagrams [J]. IEEE Transactions on Reliability, 2008, 57 (2): 336–48.
48. Shrestha A, Xing L. A logarithmic binary decision diagram-based method for multistate system analysis [J]. IEEE Transactions on Reliability, 2008, 57 (4): 595–606.
49. Rauzy A. Binary decision diagrams for reliability studies [M]. Handbook of Performability Engineering.Springer. 2008: 381–96.
50. 闵苹, 童节娟, 奚树人. 利用二元决策图求解故障树的基本事件排序 [J]. 清华大学学报 (自然科学版), 2005, 45 (12): 1646–9.
51. 张国军, 朱俊, 吴军, 等. 基于BDD的考虑共因失效的故障树可靠性分析 [J]. 华中科技大学学报 (自然科学版), 2007, 35 (9): 1–4.
52. 陶勇剑, 董德存, 任鹏. 故障树分析的二元决策图方法 [J]. 铁路计算机应用, 2009, 18 (9): 4–7.
53. 胡文军. 故障树向二元决策图的转换算法 [J]. 原子能科学技术, 2010, 44 (3): 289–93.
54. 高巍, 张琴芳. 基于二叉决策图的故障树求解法 [J]. 核技术, 2011, 34 (10): 791–5.
55. 涂序跃. 基于二元决策图的系统可靠性模块分析方法 [J]. 华东交通大学学报, 2010, 27 (5): 53–7.
56. Peng R, Zhai Q, Xing L, et al. Reliability of demand-based phased-mission systems subject to fault level coverage [J]. Reliability Engineering & System Safety, 2013, 121 18–25.
57. Mo Y, Xing L, Amari S V. A multiple-valued decision diagram based method for efficient reliability analysis of non-repairable phased-mission systems [J]. IEEE Transaction on Reliability, 2014, 63 (1): 320–30.
58. Mo Y, Xing L. An enhanced decision diagram-based method for common-cause failure analysis [J]. Proceedings of the Institution of Mechanical Engineers, Part O: Journal of Risk and Reliability, 2013, 227 (5): 557–66.
59. Xing L, Amari S V. Binary Decision Diagrams and Extensions for System Reliability Analysis [M]. John Wiley & Sons, 2015.

Chapter 3
Reliability of k-Out-Of-n Warm Standby Systems

In real-life applications, the operation of systems requires the simultaneous work of multiple units. For instance, the power station generally has multiple power plants and several power plants should operate simultaneously to meet the power requirement. Another example is the server system, which often consists of multiple servers to satisfy the assess load. To keep the system to operate at a highly reliable state, a system may be designed with some redundant subsystems or units in addition to these online units. One of the most commonly met redundant systems is the k-out-of-n system [1–3]. In this chapter, we would show how to model the reliability of k-out-of-n systems based on binary decision diagrams.

3.1 System Description

Consider a system that consists of n units $A_1, \ldots, A_n$. The system operation requires that k units should operate simultaneously. At time $t_0 = 0$, the units $A_1, \ldots, A_k$ are at the working state, while the other units are at the warm standby state. We assume that the order of switch for the warms standby units are fixed, where the warm standby unit with the lowest index, among all the warm standby units that are not failed, will be switched to the normal working state if any one of the working units fails. In such a manner, the number of working units is always k. We further assume that the switch time is negligible compared with the lifetime of units or the mission duration, i.e., the switch is done immediately. When the number of failed units exceeds $(n - k)$, in which case the number of working units will be fewer than k, then the system is deemed failed.

The lifetimes of the n units are assumed to be independent, and they can follow arbitrary distributions. Assume that the CDF of the lifetimes of the k primary working units A_i are $F_i(t)$ for $1 \leq i \leq k$, and the CDF for the lifetime of a warm standby unit under the standby state is $F_i^s(t)$ while the CDF is $F_i^o(t)$ for the unit under the normal working state for $k + 1 \leq i \leq n$.

R. Peng et al., *Reliability Modelling and Optimization of Warm Standby Systems*,
https://doi.org/10.1007/978-981-16-1792-8_3

Whenever a unit fails in the system, the system should detect and isolate this failed units. Considering the imperfect fault coverage, we use the fault level coverage model to modeling this phenomenon. In particular, we assume that the j th unit failure would be detected and isolated with a probability c_j, i.e., the coverage probability is c_j. In contrast, if a unit failure occurs but it is not correctly detected and isolated, then the failed unit will remain in the system and may cause damage to the system. For example, if a short circuit happens to a resistance but not detected, then the current in the circuit can increase and may ruin the system. Therefore, it is deemed as a system failure if a unit failure happens and not covered.

3.2 Component-Level BDD Models

Traditional BDD approaches are proper for the 0–1 cases, but the warm standby unit has an extra "warm standby" state, in addition to the "working" and "failure" states. Therefore, we need to modify the traditional BDD models to accommodate the characteristics of warm standby units.

As mentioned, the primary units $A_1, \ldots, A_k$ are at the working state at $t_0 = 0$, and any of them can fail in the following mission. Hence, the primary units only have two possible states, i.e., working or failed. Hence, a traditional BDD model can be used to describe the state of a primary unit, as illustrated in Fig. 3.1a. The terminal values of the two branches are 0 and 1, which correspond to the normal working state and the failed state, respectively. For the warm standby units, they are in the warm standby state at the very beginning and may be switched to the normal working state at some time. No matter a warm standby unit is in the standby state or the working state, it may fail at some time. Considering different scenarios, we build two BDD models for a warm standby unit, as given in Fig. 3.1b and c. Figure 3.1b represents

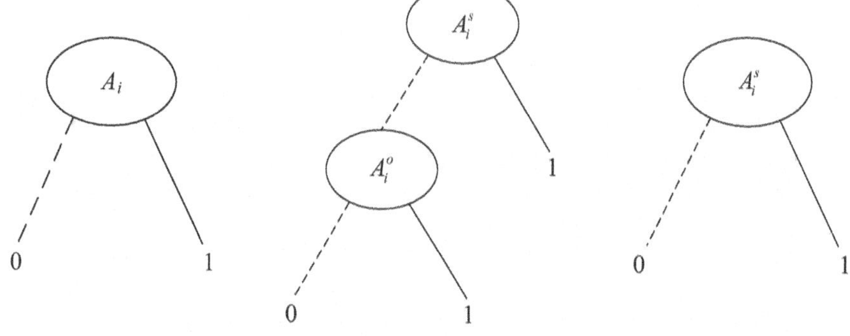

(a) Primary component (b) Warm standby unit (switched) (c) Warm standby unit (unswitched)

Fig. 3.1 Component-level BDD models for the k-out-of-n systems

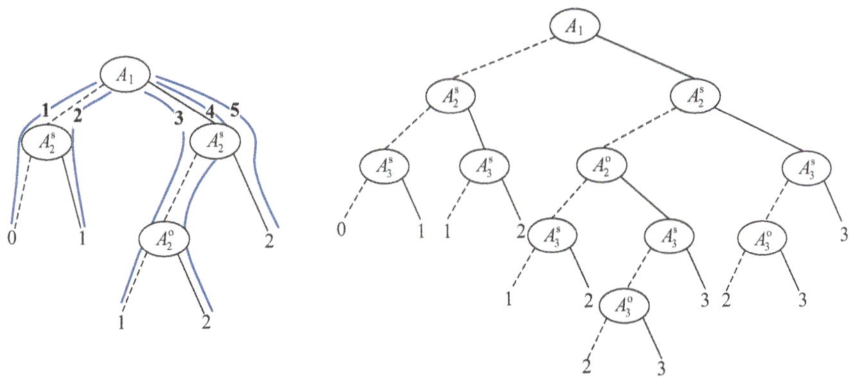

(a) System BDD of the 1-out-of-2 system

(b) System BDD of the 1-out-of-3 system

Fig. 3.2 System BDD models

the scenario that component A_i is first in the warm standby state (A_i^s), and then switched to the normal working state at some time (switched scenario). Figure 3.1b represents the case that component A_i is in the warm standby system throughout the whole mission or failed in the warm standby state (unswitched scenario).

3.3 System-Level BDD Construction

To evaluate the system reliability considering the imperfect fault coverage, we should explicitly describe the number of failed units in every possible scenario. Based on the component-level BDD model, we can iteratively construct the system-level BDD with the following steps.

Step 1. Start the BDD construction from the BDD model for unit A_1.

Step 2. Suppose that the BDD for $A_1, \ldots, A_{i-1} (2 \leq i \leq k)$ have been added to the intermediate system BDD. Denote the terminal value of a path as ξ_{i-1}. If $\xi_{i-1} \leq n - k$, then we add the BDD model for A_i under this path and we updated the terminal values of the new paths as the sum of ξ_{i-1} and the corresponding terminal values of the two branches in the component-level BDD model of A_i. Otherwise, we will not develop this path any further.

Step 3. Suppose that the BDD for $A_{i-1} (k + 1 \leq i \leq n)$ have been added to the intermediate system BDD. Denote the terminal value of a path as ξ_{i-1}. If $\xi_{i-1} \leq n - k$ and $\xi_{i-1} > i - 1 - k$, then we add the component-level BDD for A_i under the switched scenario (Fig. 3.1b to the end of this path. If $\xi_{i-1} \leq i - 1 - k$, then we add the component-level BDD for A_i under the unswitched scenario (Fig. 3.1c) to the end of this path. In these two cases, the terminal values should be updated as the sum of ξ_{i-1} and the corresponding terminal values in the component-level BDD model of A_i as in Step 2. If $\xi_{i-1} > n - k$, we will not develop this path any further.

During the system-level BDD construction, if the terminal value of a path $\xi_{i-1} > n-k$, then it indicates that the scenario corresponding to this path has more than $n-k$ failed units (note that the terminal value ξ_{i-1} actually represents the total number of component failures in a path). Therefore, the system cannot fulfil the mission requirement no matter the other units $A_i, \ldots, A_n$ fail or not. Therefore, these paths will not contribute to the system reliability, and we need not develop these paths any further. In Step 3, we also relate the terminal values to the switch of warm standby units. When $\xi_{i-1} > i-1-k$, or equivalently, $i-1-\xi_{i-1} < k$, the number of working units among the $(i-1)$ units under the scenario corresponding to this path is fewer than k. Then the warm standby units should be switched to the normal working state to replace the failed online units. Accordingly, we add the BDD in Fig. 3.1b to the end of this path. Otherwise, the warm standby needs not to be switched, and Fig. 3.1c is added to the end of this path.

3.4 System Reliability Calculation

The system-level BDD exhaustively gives all the possible scenarios that lead to a reliable system during the mission. Each path in the system BDD corresponds to a specific scenario, and the states of all the units are clearly represented. Thus, the occurrence probability of each path can be calculated accordingly based on the lifetime distributions of the units, and the system reliability is obtained as the sum of the occurrence probabilities of all the paths that lead to a reliable system.

Specifically, suppose the terminal value of a path is ξ, which means that the number of failed units in this path is ξ. Then, the probability that all the ξ failures can be correctly detected and isolated is $\prod_{r=1}^{\xi} c_r$. Denote the occurrence probability of this path conditional on all the failures is covered by $\Pr\{path\}$. Then, the probability of the path considering imperfect coverage is $\prod_{r=1}^{\xi} c_r \times \Pr\{path\}$. The system reliability considering FLC is

$$R_S(t) = \sum_{\xi \le n-k} \prod_{r=1}^{\xi} c_r \Pr\{path\}, \tag{3.1}$$

which equals to the sum of the path probabilities of these paths with no more than $n-k$ component failures. In the following, we would illustrate how the proposed BDD model for warm standby systems is applied to the system reliability evaluation.

3.5 Numerical Examples

3.5.1 *Warm Standby System with One Primary Unit and One Warm Standby Unit*

First, let us consider a warm standby system with one primary unit and one warm standby unit. Denote the lifetime distribution of the primary unit A_1 by $F_1(t)$, and the lifetime distributions of A_2 under the warm standby state and under the normal working state by $F_2^s(t)$ and $F_2^o(t)$, respectively (here we also assume that the lifetime of A_2 under the normal working state is independent of the time it has spent in the warm standby state). Based on the BDD construction procedure above, the BDD for this 1-out-of-2 warm standby system is given in Fig. 3.2a.

In the system BDD, we can notice that there are five paths in total. Specifically,

- Path 1 corresponds to the scenario that the units A_1 and A_2 survive throughout the mission, and the total number of unit failures is 0;
- Path 2 corresponds to the scenario that unit A_1 does not fail, but unit A_2 fails under the warm standby unit. The total number of unit failures is 1. Note that unit A_2 will not be switched to the normal working state if unit A_1 does not fail.
- Path 3 corresponds to the scenario that unit A_1 fails at some time during the mission, and unit A_2 is switched to the normal working state and works throughout the mission. The total number of failures is 1.
- Path 4 corresponds to the scenario that unit A_1 fails at some time during the mission, and unit A_2 is switched to the normal working state and fails after then. The total number of failures is 2.
- Path 5 corresponds to the scenario that unit A_1 fails at some time during the mission, and unit A_2 has already failed before then. The total number of failures is 2.

Therefore, the system BDD contains all the possible system scenarios, where each scenario corresponds to the combination of the states of different units. Based on the system BDD, one can enumerate all the scenarios that lead to system success.

Based on the system BDD, the system reliability can be obtained as the sum of the occurrence probabilities of all paths considering the imperfect fault coverage:

$$\begin{aligned} R_S(t) =& \sum_{\xi<2}\prod_{r=1}^{\xi} c_r \Pr\{path\} = \Pr\{path1\} + c_1(\Pr\{path2\} + \Pr\{path3\}) \\ =& R_1(t)R_2^s(t) + c_1\left(R_1(t)F_2^s(t) + \int_0^t f_1(\tau)R_2^s(\tau)R_2^o(t-\tau)d\tau\right), \end{aligned} \tag{3.2}$$

where $R_\#(\cdot) = 1 - F_\#(\cdot)$ is the reliability function of a unit, and $f_\#(\cdot)$ is the corresponding probability density function (PDF).

To improve the system reliability, we can deploy one more warm standby unit A_3 in this 1-out-of-2 system and obtain a 1-out-of-3 warm standby system. The system BDD for this new system can be obtained based on the system BDD of the 1-out-of-2 system, as shown in Fig. 3.2b.

Denote the lifetime distributions of A_3 under the warm standby state and under the normal working state by $F_3^s(t)$ and $F_3^o(t)$, respectively. The system reliability can be obtained as follows:

$$\begin{aligned} R_S(t) =& R_1(t)R_2^s(t)R_3^s(t) \\ &+c_1\begin{pmatrix} R_1(t)R_2^s(t)F_3^s(t) + R_1(t)F_2^s(t)R_3^s(t) \\ +R_3^s(t)\int_0^t f_1(\tau)R_2^s(\tau)R_2^o(t-\tau)d\tau \end{pmatrix} \\ &+c_1c_2\begin{pmatrix} R_1(t)F_2^s(t)F_3^s(t) + F_3^s(t)\int_0^t f_1(\tau)R_2^s(\tau)R_2^o(t-\tau)d\tau \\ +\int_0^t\int_0^\tau f_1(u)R_2^s(u)f_2^o(\tau-u)R_3^s(\tau)R_3(t-\tau)du\,d\tau \\ +\int_0^t f_1(\tau)F_2^s(\tau)R_3^s(\tau)R_3^o(t-\tau)d\tau \end{pmatrix} \end{aligned} \tag{3.3}$$

3.5.2 Storage System with Shared Standby Unit

Let us consider a warm standby storage system, which has a structure in Fig. 3.3 [4]. The system consists of two primary hard disks A_1, A_2 and a warm standby hard disk A_3. A_3 can be switched to the normal working state whenever a primary unit fails. If a primary unit fails, it should be immediately isolated. Otherwise, the system is deemed failed.

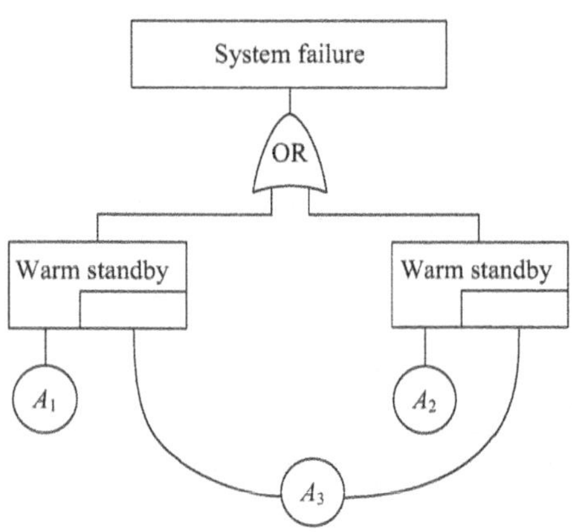

Fig. 3.3 Fault tree of the warm standby storage system with shared warm standby unit

Based on the system failure definition, we can notice that at most one unit failure is allowed in the system. In other words, the system is a 2-out-of-3 warm standby system. According to the BDD construction procedure in Sect. 3.3, we obtain the system BDD for this system, as shown in Fig. 3.4.

As the system only allows at most one component failure, the system reliability is equal to the sum of the occurrence probabilities of all the paths with terminal value no larger than 1:

$$\begin{aligned} R_S(t) &= \Pr\{path1\} + c_1(\Pr\{path2\} + \Pr\{path3\} + \Pr\{path4\}) \\ &= R_1(t)R_2(t)R_3^s(t) + c_1\left(\begin{array}{c} R_1(t)R_2(t)F_3^s(t) + R_1(t)\int_0^t f_2(\tau)R_3^s(u)R_3^o(t-\tau)d\tau \\ +R_2(t)\int_0^t f_1(\tau)R_3^s(\tau)R_3^o(t-\tau)d\tau \end{array}\right) \end{aligned} \tag{3.4}$$

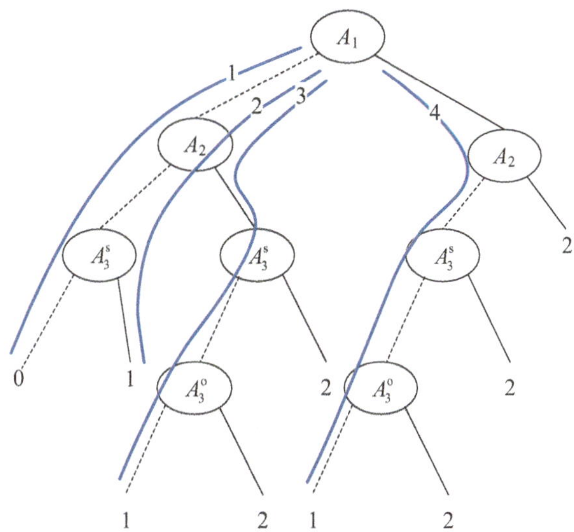

Fig. 3.4 System BDD for the warm standby storage system

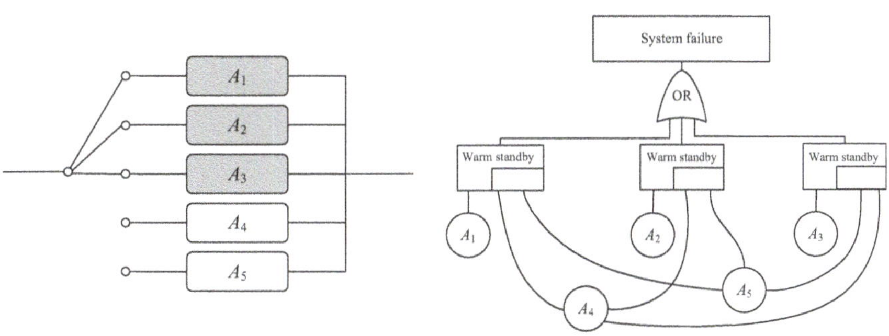

(a) System block diagram (b) Fault tree of the system

Fig. 3.5 System structure for the 3-out-of-5 warm standby system

3.5.3 Five-Unit Warm Standby System

Let us consider a slightly larger 3-out-of-5 warm standby system, as shown in Fig. 3.5. Following the same BDD construction procedure in Sect. 3.3, we can obtain the system BDD as in Fig. 3.6. To have a concise representation for the system BDD, we useG_0, G_1 and G_2 to represent three subgraphs in the system BDD. Specifically, G_0 corresponds to the states of the two standby units given that all the three primary units survive the mission. G_1 corresponds to the states of the two standby units given that one primary unit fails during the mission. G_2 corresponds to the states of the two standby units given that two primary units fail during the mission.

The system reliability can be obtained as the sum of the occurrence probabilities of the paths with terminal value smaller than 3:

$$R_S(t) = R_1(t)R_2(t)R_3(t)R_4^s(t)R_5^s(t)$$

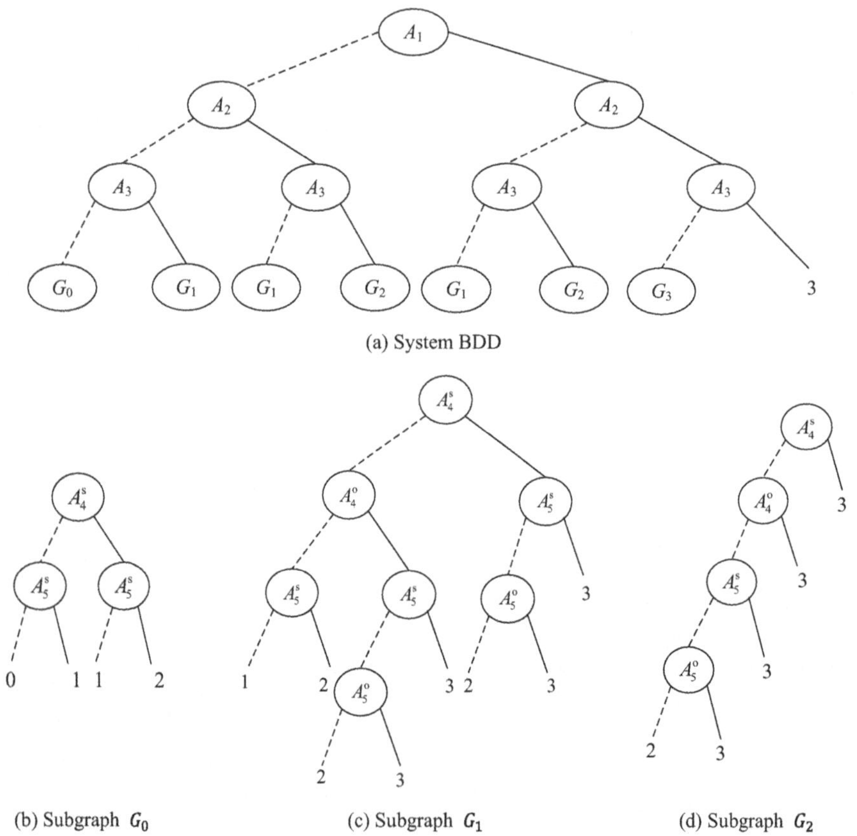

(b) Subgraph G_0 (c) Subgraph G_1 (d) Subgraph G_2

Fig. 3.6 System BDD for the 3-out-of-5 warm standby system

Table 3.1 System reliability for the 3-out-of-5 warm standby system under different coverage probabilities

t	System reliability		
	$c_1 = c_2 = 1$	$c_1 = c_2 = 0.9$	$c_1 = 0.9, c_2 = 0.8$
30	0.9808	0.9239	0.8692
50	0.8553	0.7655	0.6813
70	0.5994	0.5171	0.4412

$$
\begin{aligned}
&+c_1\begin{pmatrix} R_1R_2(t)R_3(t)\sum\limits_{i,j}^{\{1,2\}_P} F_i^s(t)R_j^s(t) \\ +R_5^s(t)\sum\limits_{l,m,n}^{\{1,2,3\}_C} R_l(t)R_m(t)\int_0^t R_4^s(\tau)R_4^o(t-\tau)dF_n(\tau) \end{pmatrix} \\
&+c_1c_2\begin{pmatrix} R_1(t)R_2(t)R_3(t)F_4^s(t)F_5^s(t) \\ +\sum\limits_{\{l,m,n\}}^{\{1,2,3\}_C} R_l(t)R_m(t)\int_0^t F_4^s(\tau)R_5^s(\tau)R_5^o(t-\tau)dF_n(\tau) \\ +\sum\limits_{\{l,m,n\}}^{\{1,2,3\}_C} R_l(t)R_m(t)\int_0^t R_5^s(\tau)R_5^o(t-\tau)\int_0^\tau R_4^s(u)f_4^o(\tau-u)f_n(u)dud\tau \\ +F_5^o(t)\sum\limits_{\{l,m,n\}}^{C_{1,2,3}} R_l(t)R_m(t)\int_0^t R_4^s(u)R_4^o(t-\tau)dF_n(\tau) \\ +\sum\limits_{\{l,m,n\}}^{\{1,2,3\}_P} R_l(t)\int_0^t R_4^s(\tau)R_4^o(t-\tau)f_m(\tau)\int_\tau^t R_5^s(u)R_5^o(t-u)f_n(u)dud\tau \end{pmatrix}
\end{aligned}
\quad (3.5)
$$

where,

$\{1, 2\}_P = \{\{1, 2\}, \{2, 1\}\}$,

$\{1, 2, 3\}_C = \{\{1, 2, 3\}, \{2, 3, 1\}, \{3, 1, 2\}\}$,

$\{1, 2, 3\}_P = \{\{1, 2, 3\}, \{1, 3, 2\}, \{2, 3, 1\}, \{2, 1, 3\}, \{3, 1, 2\}, \{3, 2, 1\}\}$.

Assume that the lifetime distributions of the primary units A_1, A_2 and A_3 independently follows the Weibull distribution $R_{\text{prim}}(t) = \exp\{-(t/\eta)^m\}$ with $m = 1.5$, $\eta = 100$. The warm standby units A_4 and A_5 follow a Weibull distribution $R^{\text{s}}_{\text{standby}}(t) = \exp\{-(t/100)^2\}$ under the warm standby state, and $R^{\text{o}}_{\text{standby}}(t) = \exp\{-(t/50)^2\}$ under the normal working state. The system reliability for this system under different coverage probabilities is listed in Table 3.1. Clearly, the system reliability is significantly dependent on the coverage probabilities, implying the importance of imperfect fault coverage phenomenon.

References

1. She J, Pecht M G. Reliability of a k-out-of-n warm-standby system [J]. Reliability IEEE Transactions on, 1992, 41 (1):72-75.
2. Myers A F. *k*-out-of-*n*:G system reliability with imperfect fault coverage [J]. IEEE Transactions on Reliability, 2007, 56 (3):464-473
3. Amari S V, Pham H, Misra R B. Reliability characteristics of *k*-out-of-*n* warm standby systems [J]. IEEE Transactions on Reliability, 2012, 61 (4):1007-1018

4. Tannous O, Xing L, Dugan J. Reliability analysis of warm standby systems using sequential BDD [A]. Annual Reliability and Maintainability Symposium (RAMS2011), Lake Buena Vista, Florida, USA, 2011 [C]. Piscataway, NJ: IEEE, 1–7.

Chapter 4
Reliability of Demand-Based Warm Standby Systems

In Chap. 3 we present a BDD-based approach for the reliability modeling of k-out-of-n wamr standby systems subject to imperfect fault coverage. This approach enumerates the possible combinations of the component states that can result in the system success, and obtain the system reliability as the sum of the occurrence probabilities of these combinations. The advantage of this approach is that the BDD representation is relatively simple, but the calculation of the occurrence probability is complicated because a certain combination of component states may correspond to several different scenarios in the system operation. Consequently, one has to further analyze the scenarios embedded by each combination of component states, which complicates the system reliability evaluation. To address this problem, this Chapter introduces an MDD-based method for reliability modeling of more general demand-based warm standby systems. The MDD-based method directly models the failure sequences in the system and leads to a clearer system reliability evaluation approach.

4.1 System Description

We consider demand-based warm standby systems in this Chapter, which can be viewed as a generalization of the k-out-of-n systems. The system consists of n components $A_1, \ldots, A_n$, and the failures of all the components are independent. We assume that the lifetime of each component can follow an arbitrary distribution under the normal working state or the warm standby state. The lifetime of a warm standby component A_i can be modeled as outlined in Sect. 4.1. To account for the time dependent failure behavior of the components in the warm standby system, different lifetime distributions are utilized before and after the standby component replacing the online failed component. Specifically, component A_i is assumed to follow the lifetime distribution $F_{A_i^s}(t)$ at the standby state and $F_{A_i^o}(t)$ at the fully powered up state. Note that $F_{A_i^s}(t)$ and $F_{A_i^o}(t)$ can be of the same type, but with different parameter values.

R. Peng et al., *Reliability Modelling and Optimization of Warm Standby Systems*,
https://doi.org/10.1007/978-981-16-1792-8_4

Assume that $A_i (1 \leq i \leq n)$ has a nominal capacity $w_i > 0$ under the normal working state, and the system capacity equals the sum of the capacities of all the working components. Suppose that the system capacity has to meet a prefixed demand d, and the system is deemed failed if the system capacity is smaller than the demand. Note that if all the components have the same nominal capacity $w_i = w$, then the system can function properly only if the number of the working components is no smaller than $k = [d/w]$. Here, $[x]$ denotes the smallest interger no smaller than x. As such, the demand-based system is just an k-out-of-n system.

In the beginning, k components $A_1, A_2, \ldots, A_k$ are at the online state and the remaining $(n - k)$ components $A_{k+1}, \ldots, A_n$ are at the standby state. During the mission, a standby component can be activated to become online. For clarity, we use superscripts "o" and "s" to distinguish the online state and the standby state for $A_{k+1}, \ldots, A_n$, i.e., A_i^o indicates component A_i is in the online state and A_i^s indicates the standby state, $(k + 1) \leq i \leq n$. Whenever an online component fails, one or more standby components will be powered up to ensure that the total capacity of the working components is not smaller than the required system demand. The order to be powered up depends on the index of the standby components. Specifically, component A_j will not be powered up if component A_i is still working in warm standby mode for $k < i < j \leq n$.

We also consider the imperfect fault coverage effect, and use FLC model for the fault coverage. Assume that the first fault would be covered with probability c_1, the second fault would be covered with probability c_2 and so on. The system will fail if there are any uncovered faults.

4.2 Decision Diagram Based on Failure Sequences

According to the system description given in Sect. 4.1, three aspects have to be considered when analyzing the system reliability. First, since the warm standby components may fail either in the standby state or under the fully powered up state after replacing the online failed component, these two states should be handled in a different manner. Second, as the system suffers the FLC, the number of failed components in the system has to be incorporated into the MDD representation. Third, the capacity of each component should also be reflected in the MDD representation to determine the state of the system. In the light of these considerations, we extend the traditional MDD model to analyze the reliability of the demand-based warm standby systems subject to FLC. In the the proposed MDD-based approach, instead of using the nodes in the MDD to model components, the nodes are used to model the failure sequence.

First, we show how to build the MDD representation for a single failure. Consider the first failure in the system. Since the system has n components and any one of the component can fail as the first failure, we denote it by an MDD node with $(n + 1)$ branches, where the ith $(1 \leq i \leq n)$ branch represents the failure of component A_i

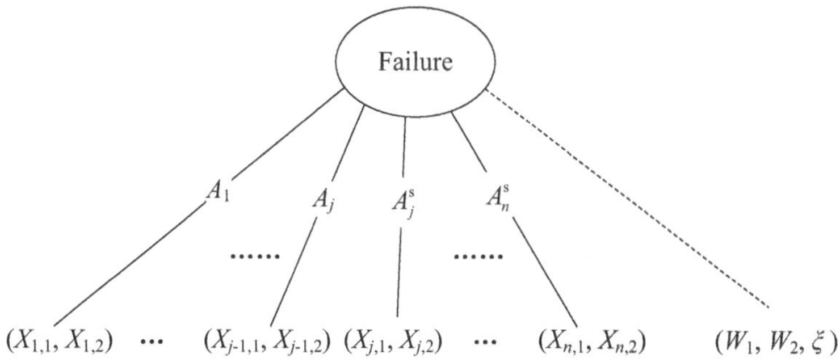

Fig. 4.1 MDD representation of a single failure

and the $(n + 1)$th branch denotes that there is no first failure (i.e. no failure happens at all). An illustration for the MDD is given in Fig. 4.1.

The terminal value of branch $i(1 \leq i \leq n)$ is denoted by a bivariate $\left(X_{i,1}, X_{i,2}\right)$, where $X_{i,1}$ and $X_{i,2}$ denote the actual system capacity loss and the available system capacity loss due to the failure of component A_i, respectively. Here, the available system capacity refers to the sum of the capacities of all the available components, no matter they are operational or in the warm standby state. If A_i is online, both $X_{i,1}$ and $X_{i,2}$ equal to its nominal capacity, since the failure of an online component will reduce both the actual system capacity and the available system capacity. On the contrary, if A_i is a standby component, the corresponding $X_{i,1}$ equals to 0 because it does not contribute to the system functioning in the standby state and thus its failure will not reduce the actual system capacity, whereas $X_{i,2}$ equals to w_i because its failure will reduce the available capacity of the system.

The terminal value of the rightmost $(n + 1)$th branch is denoted by a triple (W_1, W_2, ξ), where W_1 is the actual system capacity, W_2 is the available system capacity and ξ is the number of component failures given that the first failure does not happen. Clearly, for the MDD of the first failure $W_1 = \sum\limits_{i=1}^{k} w_i$, $W_2 = \sum\limits_{i=1}^{n} w_i$ and $\xi = 0$. For each branch except for the rightmost one, calculate $W_2 - X_{i,2}$: if the value is less than the demand, which indicates that the system will fail if the failure corresponding to this branch happens, then remove this branch. Figure 4.2 illustrates the MDD representation for the First Failure (FF) before removing any branches.

The MDD representation for the second failure is modified from the MDD of the first failure. Specifically, corresponding to a certain branch $i(1 \leq i \leq n)$ of the first MDD, the MDD for the second failure is built using the following four-step procedure.

Step-1: Remove branch i. Since this branch denotes the failure of component A_i as the first failure, it cannot fail as the second failure again.

Step-2: Recalculate $W_1 = W_1 - X_{i,1}$, $W_2 = W_2 - X_{i,2}$. If W_1 is smaller than the demand, which means the current system capacity cannot meet the demand and some standby components should be fully powered up, then the terminal value

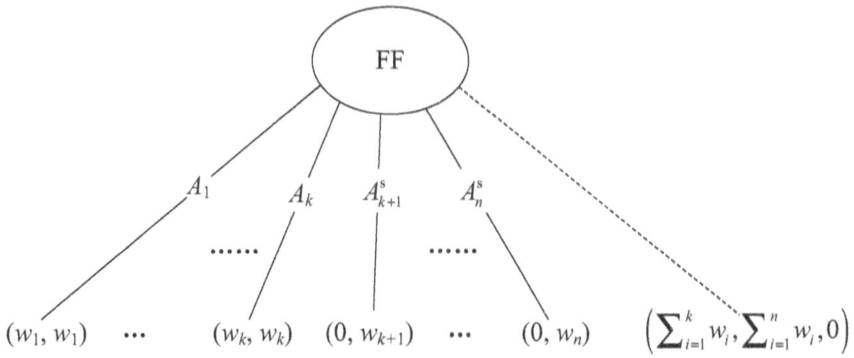

Fig. 4.2 MDD representation for the first failure in the system

$X_{i,1}$ of some branches is changed from 0 to the corresponding nominal capacity w_i in sequence (from the component with lower index). Meanwhile, W_1 is updated by adding the newly changed $X_{i,1}$ until W_1 is not smaller than the demand. This procedure corresponds to the failure dynamics that an online component A_i fails and is replaced by some standby components $\left(A^s_{j1}, \ldots, A^s_{jr}\right)$, $1 \leq r \leq (n-k)$. To explicitly denote this procedure, we denote the node as "$A_i \rightarrow \left(A^s_{j1}, \ldots, A^s_{jr}\right)$". On the other hand, if W_1 is not smaller than the demand (for instance, the first failed component is a standby component A^s_i whose failure will not reduce the actual system capacity; or the first failed component is an online component A_i but the remaining actual system capacity can still meet the demand), then no standby components is required to be activated. In this case, denote the node as "A^s_i" or "A_i" to indicate the first failed component.

Step-3: For each branch in the MDD of the second failure, calculate $W_2 - X_{i,2}$: if it is smaller than the demand, then remove this branch.

Step-4: Change the terminal value ξ of the rightmost branch from 0 to 1 to indicate that if no second failure happens, there would be only one component failure in the system.

Figure 4.3a illustrates the MDD for the second failure given that the first failed component is A_1 and the standby component A_{k+1} is fully powered up to replace A_1. Note that the terminal value for A_{k+1} changes to (w_{k+1}, w_{k+1}) and the terminal value for the rightmost branch changes to $W_1 = \sum_{i=2}^{k+1} w_i$, $W_2 = \sum_{i=2}^{n} w_i$, $\xi = 1$. Figure 4.3b illustrates the MDD for the second failure given that the first failed component is the standby component A_{k+1}, where the actual system capacity is not reduced and no standby components is fully powered up. For the terminal value of the rightmost branch, W_1 is kept unchanged while W_2 is changed to $\sum_{i=1, i \neq k+1}^{n} w_i$.

Similarly, the MDD representation for the $(l+1)$th$(2 \leq l \leq n)$ failure can be developed from the MDD of the lth failure using the above four-step modification

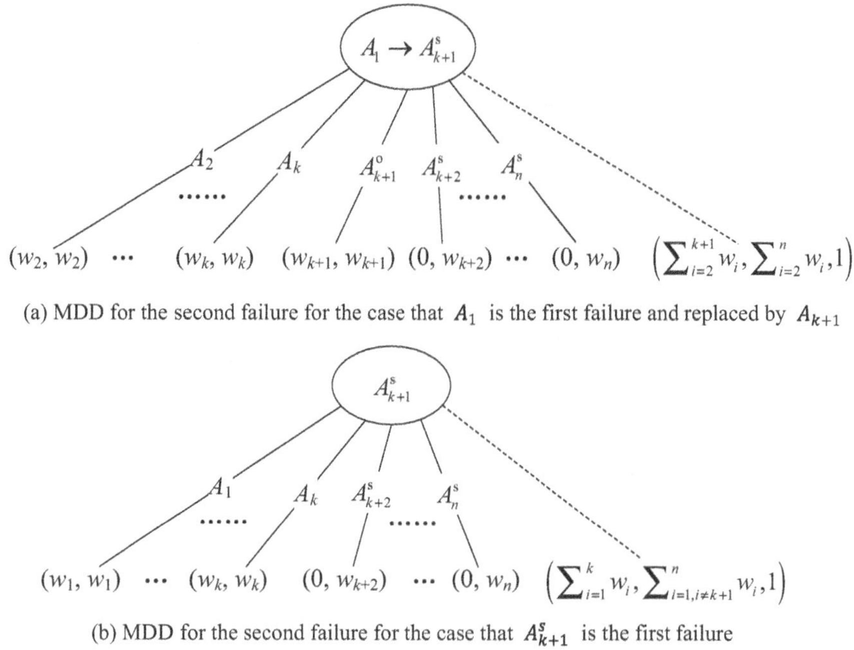

(a) MDD for the second failure for the case that A_1 is the first failure and replaced by A_{k+1}

(b) MDD for the second failure for the case that A^s_{k+1} is the first failure

Fig. 4.3 MDD representation for the second failure in the system

procedure. Figure 4.4 illustates the iterative way to obtain the MDD representation for the $(l+1)$th failure from the MDD for the lth failure.

4.3 Construction of System-Level MDD

Based on the MDD of each individual failure, the system MDD can be constructed in an iterative way:

1. Build the MDD representation of the first failure, i.e., $j = 1$;
2. Build the MDD representation for the $(j+1)$th failure based on the MDD representation of the jth failure and add it to the corresponding terminal of the constructed system MDD. If the MDD for the $(j+1)$th failure has only one branch left, i.e. the rightmost branch indicating no $(j+1)$th failure happens, then use ξ as the terminal value instead of the triple (W_1, W_2, ξ) for simplicity;
3. $j = j + 1$;
4. If $j \geq n$ or no $(j+1)$th failure happens on any branch of the MDD of any jth failure, then stop and the system MDD is obtained; otherwise, turn to Step 2.

Consider a DB-WSS with two online components A_1, A_2 and one standby component A_3 at the beginning of the mission. The capacities of A_1, A_2 and A_3 are 1, 2

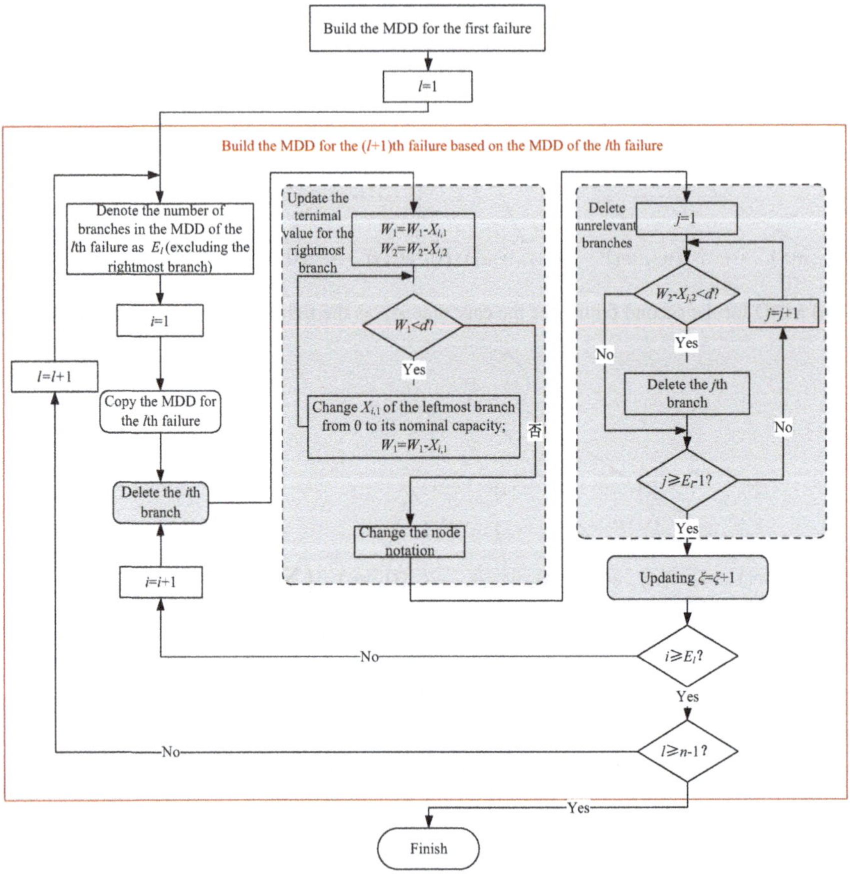

Fig. 4.4 Flow chart to iteratively build MDD for the failures in the system

and 3, respectively. The system demand is $d = 3$. Then the MDD representations for each failure and the system MDD are given in Fig. 4.5.

4.4 Derivation of System Reliability

Observing the constructed MDD for the example DB-MSS in Fig. 4.5g, the leftmost path "{FF} $\rightarrow \{A_1 \rightarrow A_3^s\} \rightarrow \{A_2\} \rightarrow \{2\}$" means that: the first failure is A_1 and replaced by A_3^s and A_2 fails as the second failure; the number of the component failures in the path is 2 and the system survives during the mission time. Similarly, path "{FF} $\rightarrow \{A_3^s\} \rightarrow \{1\}$" indicates that the first failure is component A_3 in the standby state: there is only one component failure in this path and the system survives.

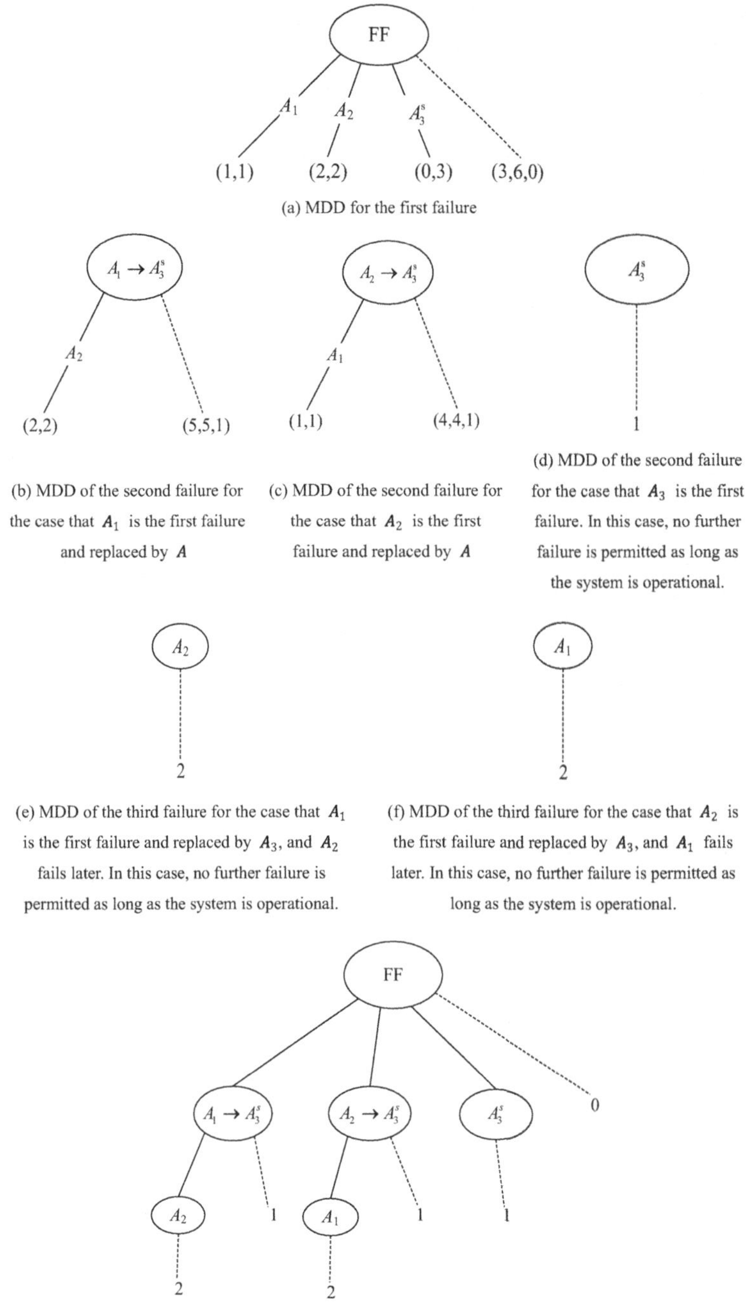

Fig. 4.5 Illustration of the MDD construction proedure for a three-component system

Therefore, the MDD of a DB-WSS models all the possible failure sequences that contribute to the system functioning.

4.4.1 Probability of Edge in System MDD

In order to obtain the system reliability, it is essential to investigate the occurrence probability of each path in the constructed MDD. Again, consider the leftmost path in Fig. 4.5g "{FF} $\rightarrow$ $\{A_1 \rightarrow A_3^s\}$ $\rightarrow$ $\{A_2\}$ $\rightarrow$ {2}" for instance. Taking the FLC into consideration, the occurrence probability of this path is

$$\begin{aligned} \Pr(T) &= c_1 c_2 \int_{t_0}^{T} f_1(t_1) R_3^s(t_1) R_3^o(T - t_1) dt_1 \cdot (F_2(T) - F_2(t_1)) \\ &= R_0(T) \times \frac{\int_{t_0}^{T} R_3^s(t_1) R_3^o(T - t_1) dF_1(t_1)}{R_1(T) R_3^s(T)} \times \frac{\int_{t_1}^{T} dF_2(t_2)}{R_2(T)} \times c_1 c_2 \end{aligned} \tag{4.1}$$

where T is the mission time, $t_0 = 0$ is the beginning time, $f_*(\cdot)$ is the PDF, $F_*(\cdot)$ is the CDF and $R_*(\cdot) = 1 - F_*(\cdot)$ is the reliability function. In the last equality, the occurence probability of the path is decomposed into four terms. The first term is a factor defined as $R_0(T) = R_1(T)R_2(T)R_3^s(T)$, the second term corresponds to the edge leading to node "$A_1 \rightarrow A_3^s$", the third term corresponds to the edge leading to node "A_2" and the last term corresponding to the edge leading to the sink node "2". Hence, the occurrence probability of a certain path can be expressed as the product of the occurrence probabilities of the edges constituting the path and the factor $R_0(T)$.

It is noticed that the occurrence probability expression corresponding to each edge in the MDD is characterized by four elements: the failed component, the standby components replacing the failed component, the starting time (the time to be fully powered up) of the failed component and the failure time of the failed component. For the failure denoted by "A_i", "A_i^s" or "$A_i \rightarrow \left(A_{j1}^s, \ldots, A_{jr}^s\right)$", the starting time of the failed component is $t_0 = 0$, hence, only the failure time is needed to be specified. For the failure "A_i^o" or "$A_i^o \rightarrow \left(A_{j1}^s, \ldots, A_{jr}^s\right)$", both the starting time t_{l_1} and the failure time t_{l_2} are required to describe the failure. Each failure will introduce a new failure time t_l which is the integral variable in the corresponding probability expressions and located between the failure time of the last failure and the mission time T. For instance, corresponding to the failure "$A_1 \rightarrow A_3^s$", term $\int_{t_0}^{T} R_3^s(t_1) R_3^o(T - t_1) dF_1(t_1)$ contains a new introduced integral variable t_1 indicating the failure time of A_1 and located in (t_0, T). Similarly, term $\int_{t_1}^{T} dF_2(t_2)$ introduces t_2, where t_2 is located in (t_1, T) indicating that A_2 fails after A_1. With this observation, the failure descriptions are slightly modified to incorporate the related time and generic probability expressions for different edges in the MDD are given as follows.

- For failure $A_i \to \left(A^s_{j1}, \ldots A^s_{jr}\right)$, suppose the failure time introduced by the last failure is t_{l-1}, then the new introduced failure time by A_i is t_l. Re-express "$A_i \to \left(A^s_{j1}, \ldots, A^s_{jr}\right)$" as "$A_i \to \left(A^s_{j1}, \ldots, A^s_{jr}\right), t_l$" to explicitly represent the failure time of A_i and the occurrence probability of the corresponding edge is:

$$\frac{\int_{t_{l-1}}^{T} \prod_{v=1}^{r} \left[R^s_{j_v}(t_l) R^o_{j_v}(T - t_l) \right] dF_i(t_l)}{\prod_{v=1}^{r} R^s_{j_v}(T) \times R_i(T)} \tag{4.2}$$

- For failure "A_i" or "A^s_i", suppose the failure time introduced by the last failure is t_{l-1}, then the failure can be re-expressed as "A_i, t_l" or "A^s_i, t_l" to indicate that component A_i or A^s_i fails at $t_l \in (t_{l-1}, T)$. The occurrence probability of the edge leading to node "A_i, t_l" or "A^s_i, t_l" is:

$$\frac{\int_{t_{l-1}}^{T} dF_i(t_l)}{R_i(T)} or \frac{\int_{t_{l-1}}^{T} dF^s_i(t_l)}{R^s_i(T)} \tag{4.3}$$

- As mentioned, failure "A^o_i" needs both the starting time t_{l_1} and its failure time t_{l_2} to characterize its behavior. Therefore, the failure is re-expressed as "A^o_i, t_{l_1}, t_{l_2}" and the occurrence probability of the corresponding edge is:

$$\frac{\int_{t_{l_2-1}}^{T} dF^o_i\left(t_{l_2} - t_{l_1}\right)}{R^o_i\left(T - t_{l_1}\right)} \tag{4.4}$$

- Failure "$A^o_i \to \left(A^s_{j1}, \ldots, A^s_{jr}\right)$" also needs two time indices: the starting time t_{l_1} and the failure time t_{l_2} to describe its behavior. So, this failure can be re-expressed as "$A^o_i \to \left(A^s_{j1}, \ldots, A^s_{jr}\right), t_{l_1}, t_{l_2}$" and the occurrence probability of the corresponding edge is:

$$\frac{\int_{t_{l_2-1}}^{T} \prod_{v=1}^{r} \left[R^s_{j_v}\left(t_{l_2}\right) R^o_{j_v}\left(T - t_{l_2}\right) \right] dF^o_i\left(t_{l_2} - t_{l_1}\right)}{\prod_{v=1}^{r} R^s_{j_v}(T) \times R^o_i\left(T - t_{l_1}\right)} \tag{4.5}$$

In fact, all the nodes can be represented by in a standard form as follows:

$$A^*_i \to \left(A^s_{j1}, \ldots, A^s_{jr}\right), \left(t_{l_1}, t_{l_2}\right], \tag{4.6}$$

where the superscript "*" can be "s", "o" or empty, representing the standby state for warm standby components, the operational state for warm standby components and the operational state for original online components. The index r can be zero, in which case no warm standby component is switched on. $\left(t_{l_1}, t_{l_2}\right]$ indicates the interval that A_i^* is operational, with t_{l_1} being the time that A_i^* begins to work at state indicated by "*", and t_{l_2} being the time that A_i^* is fails in the state "*". t_{l_1} can be $t_0 = 0$. Recall that this node corresponds to the event indicated by the edge leading to this node. The associated probability for the edge leading to this node is:

$$\frac{\int_{t_{l_2-1}}^{T} \prod_{v=1}^{r}\left[R_{jv}^{s}(t_{l_2}) R_{jv}^{o}(T - t_{l_2})\right] dF_i^*(t_{l_2} - t_{l_1})}{\prod_{v=1}^{r} R_{jv}^{s}(T) \times R_i^*(T - t_{l_1})}. \tag{4.7}$$

In this expression, the numerator corresponds to the occurrence probability of the following event: A_i begins working at t_{l_1} and fails at $t_{l_2} \in \left(t_{l_2-1}, T\right]$; meanwhile, r warm standby components $A_{j1}, \ldots, A_{jr}$ are switched to the online state. The denominator corresponds to the occurrence probability that A_i and the warm standby components $A_{j1}, \ldots, A_{jr}$ survive until T.

During the construction of the component MDD, it is also convenient to update the notation for the working interval $\left(t_{l_1}, t_{l_2}\right]$. The following two-step operation should be incorporated in Step-2 of the MDD construction procedure in Sect. 4.2:

Step-2.1: If the failed component is a warm standby component in the normal working state, then we have to tranverse upwards throught the path to find the time when it is switched to the online state, i.e. t_{l_1}. This is just the failure time of the particular online component replaced by this warm standby component. Otherwise, just let $t_{l_1} = t_0$.

Step-2.2: Update the failure time t_{l_2} as t_ξ.

To complete the assignment, the edge leading to the node "ξ" is assigned the probability $\prod_{j=1}^{\xi} c_j$ to indicate the coverage factor of this path. Define the factor $R_0(T) = \prod_{i=1}^{k} R_i(T) \times \prod_{i=k+1}^{n} R_i^s(T)$. Then, the occurrence probability of a certain path is equal to the product of the probabilities of all the edges in the path timing the factor $R_0(T)$ and the system reliability is equal to the occurrence probabilities of all the paths.

For illustration, Fig. 4.6 gives the probability of each edge in Fig. 4.5(g) and the system reliability can be easily obtained as:

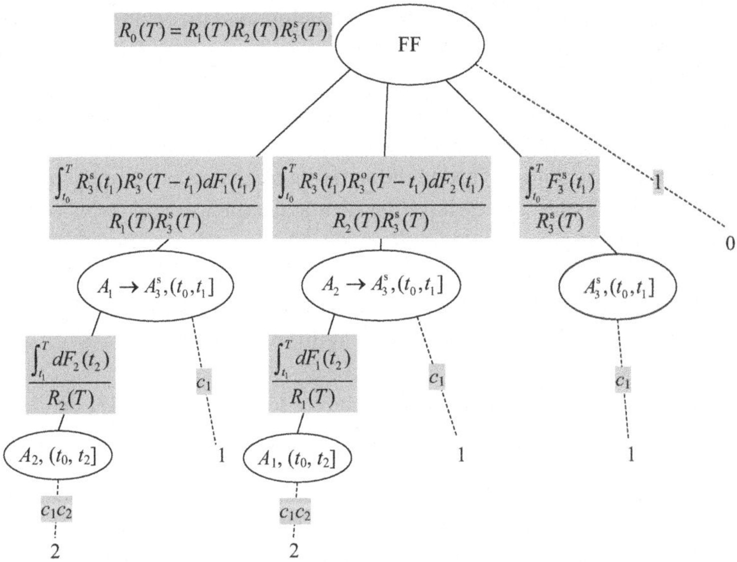

Fig. 4.6 Occurrence probabilities of all the edges in the example MDD

$$
\begin{aligned}
R_S(T) &= R_1(T)R_2(T)R_3^s(T) \\
&\times \left(\begin{array}{l} \frac{\int_{t_0}^{T} R_3^s(t_1)R_3^o(T-t_1)dF_1(t_1)}{R_1(T)R_3^s(T)} \times \left(c_1c_2 \frac{\int_{t_1}^{T} dF_2(t_2)}{R_2(T)} + c_1 \right) \\ + \frac{\int_{t_0}^{T} R_3^s(t_1)R_3^o(T-t_1)dF_2(t_1)}{R_2(T)R_3^s(T)} \times \left(c_1c_2 \frac{\int_{t_1}^{T} dF_1(t_2)}{R_1(T)} + c_1 \right) \\ + c_1 \frac{\int_{t_0}^{T} dF_3^s(t_1)}{R_3^s(T)} + 1 \end{array} \right) \\
&= R_1(T)R_2(T)R_3^s(T) + c_1 \left(\begin{array}{c} R_2(T)\int_{t_0}^{T} R_3^s(t_1)R_3^o(T-t_1)dF_1(t_1) \\ +R_1(T)\int_{t_0}^{T} R_3^s(t_1)R_3^o(T-t_1)dF_2(t_1) \\ +R_1(T)R_2(T)F_3^s(T) \end{array} \right) \\
&+ c_1c_2 \left(\begin{array}{l} \int_{t_0}^{T} R_3^s(t_1)R_3^o(T-t_1)dF_1(t_1)\int_{t_1}^{T} dF_2(t_2) \\ + \int_{t_0}^{T} R_3^s(t_1)R_3^o(T-t_1)dF_2(t_1)\int_{t_1}^{T} dF_1(t_2) \end{array} \right)
\end{aligned} \tag{4.8}
$$

4.4.2 Bottom-Up System Reliability Calculation

From the system reliability evaluation point of view, the time indices related to each failure should be incorporated into the MDD. Actually, it is convenient to incorporate the time indices in the MDD by using both the failure and the related time indices to define the node when building the MDD representation for each failure, as shown in Fig. 4.6.

After the MDD is constructed, it can be further reduced by merging isomorphic sub-trees: if the subtrees of two nodes have identical structure, then one of the subtrees can be removed and both nodes lead to the same subtree. We shall emphasize that two nodes are isomorphic only if they have the identical expression in terms of the standard form in (4.6). As a counterexample, node "$A_2, (t_0, t_1]$" and node "$A_2, (t_0, t_2]$" cannot be treated as isomorphic.

After the MDD is reduced, the system reliability can be obtained in an upwards manner. Let the cumulative occurrence probability of a certain node be:

$$P_C(\text{node } i) = \sum_{j \text{ s the child of node } i} \Pr(\text{edge leading to node } j) \times P_C(\text{node } j) \quad (4.9)$$

The cumulative occurrence probabilities of sink nodes (i.e. nodes indicating the number of failures in the path) are defined as 1. Then the system reliability is equal to the cumulative occurrence probability of the top-most node multiplied by the factor $R_0(T)$. For example, the cumulative occurrence probability of each node in the example three-component system in Fig. 4.5 after reduction is highlighted in Fig. 4.7. It is clear that the system reliability is equal to $R_0(T) \times P_C(topmost\ node)$, which is identical to (4.8).

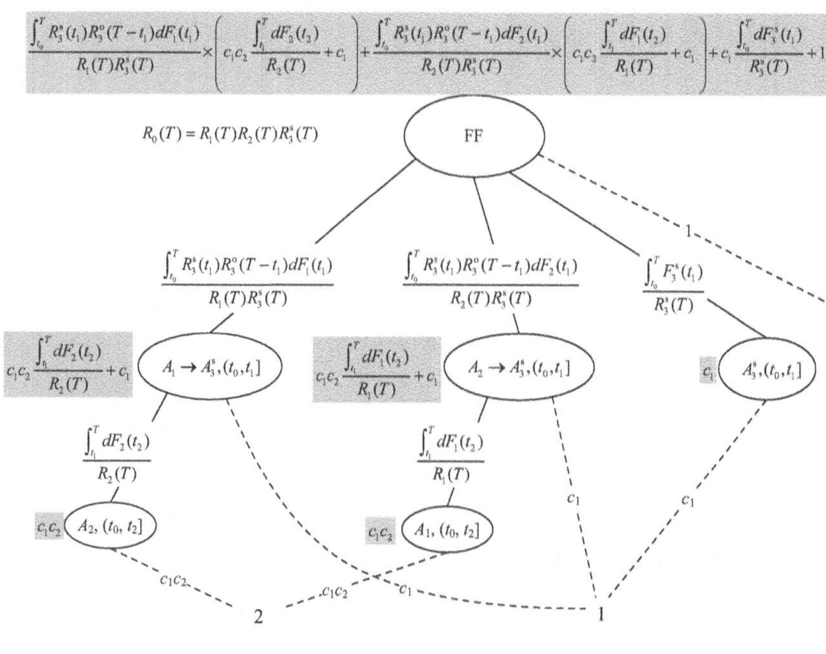

Fig. 4.7 Reduced MDD and the cumulative occurrence probabilities of the edges

4.4.3 Scale of the System MDD

The computation effort consumed by the proposed algorithm mainly results from the MDD construction procedure. According to the proposed construction algorithm, the MDD denoting the lth$(1 \leq l \leq n-1)$ failure has no more than $\prod\limits_{j=n-l+1}^{n} j$ terminals excluding terminals for "no lth failure". Taking the top-most node into account, the number of non-sink nodes in the MDD is no more than $1 + \sum\limits_{l=1}^{n-1}\left(\prod\limits_{j=n-l+1}^{n} j\right) = n! \times (e-1)$, where $n! \times (e-1)$ denotes the greatest integer less than $n! \times (e-1)$. For example, the number of non-sink nodes in the example warm standby system with 2 online components and 1 standby component is 6, whereas the upper bound is $3! \times (e-1) = 10$. The computation effort for system reliability calculation with the upwards algorithm (4.9) is almost linear with the number of the non-sink nodes, hence, the computational effort required to obtain the explicit system reliability expression is proportional to $n! \times (e-1)$. To obtain the value of the system reliability at a given time T, additional computation effort is required to solve the integrals involved in the system reliability expression.

4.5 Numerical Calculation of System Reliability

According to the preceding description, the occurrence probability of each path in the MDD can be represented by the following form:

$$\Pr\{Path\} = R_0(T) \times \prod_{j=1}^{\xi} c_j \times \int_{\Delta_T^{\xi}} g_{path}(t_1, \ldots, t_\xi) dt_1 \ldots dt_\xi, \tag{4.10}$$

where ξ denotes the number of failed components in the path, T is the mission time, and $\int_{\Delta_T^{\xi}} g_{path}(t_1, \ldots, t_\xi) dt_1 \ldots dt_\xi$ is the occurrence probability of the path excluding $R_0(T)$ and the FLC effect. Note that every component failure will introduce a failure time. If a path corresponds to ξ component failures, then ξ failure times $t_1, \ldots, t_\xi$ will be introduced. The failure times $t_1, \ldots, t_\xi$ are random, and they should satisfy $0 < t_1 < \ldots < t_\xi < T$. Therefore, the integration region is a ξ-simplex $\Delta_T^{\xi} = \{(t_1, \ldots, t_\xi) | 0 < t_1 < \ldots < t_\xi < T\}$.

The ξ-dimensional integrand $g_{path}(t_1, \ldots, t_\xi)$ can be viewed as a probability density, which can be obtained from the probability of edges in the path. For example, the occurrence probability of the leftmost path in Fig. 4.5g, which is given in (4.1), can be reformulated as

$$\Pr\{Path\} = R_0(T) \times c_1 c_2 \times \int_{t_0}^{T}\int_{t_1}^{T} \frac{R_3^s(t_1)R_3^o(T-t_1)f_1(t_1)}{R_3^s(T)R_1(T)} \times \frac{f_2(t_2)}{R_2(T)} dt_1 dt_2. \tag{4.11}$$

This indicates that the integrand is

$$g_{path}(t_1, t_2) = \frac{R_3^s(t_1)R_3^o(T-t_1)f_1(t_1)}{R_3^s(T)R_1(T)} \times \frac{f_2(t_2)}{R_2(T)}.$$

where the integration region is a 2-simplex, i.e., a triangle $\Delta_T^2 = \{(t_1, t_2)|0 < t_1 < t_2 < T\}$.

The difference of this expression in terms of a ξ-dimensional intergration with the expression derived from the standard form in (4.7) is that the expression in (4.10) explicitly defines the integration region and the integrand. With this form of expression, the total occurrence probability of all the paths that lead to the terminal "ξ" can be expressed as

$$R_0(T) \times \prod_{j=1}^{\xi} c_j \times \int_{\Delta_T^{\xi}} G_\xi\left(t_1, \ldots, t_\xi\right) dt_1 \ldots dt_\xi, \tag{4.12}$$

where

$$G_\xi\left(t_1, \ldots, t_\xi\right) = \sum_{\text{path leading to } \xi} g_{\text{path}}\left(t_1, \ldots, t_\xi\right)$$

is the "probability density" of the paths that lead to the terminal "ξ". If the system can tolerate at most N_f component failures without shutdown, then the system reliability can be represented as

$$R_S(T) = R_0(T) \times \left(1 + \sum_{\xi=1}^{N_f} \prod_{j=1}^{\xi} c_j \times \int_{\Delta_T^{\xi}} G_\xi\left(t_1, \ldots, t_\xi\right) dt_1 \ldots dt_\xi\right), \tag{4.13}$$

where the path leading to the terminal "$\xi = 0$" occurs with probability $R_0(T)$. In (4.13), the system reliability is formulated as a sum of integrals. Each integral corresponds to the occurrence probability of those paths that lead to the same terminal ξ, which has a determined integration region, i.e., a ξ-simplex, and the integrand is relevant to the corresponding paths. If the values of the integrand in the ξ-simplex can be numerircally obtained, then numerical integration methods can be applied to calculate the system reliability at a given mission time T.

4.5.1 Decomposition of System MDD

To calculate the system reliability using numerical methods, we need to evaluate the values of the integrand $G_\xi(t_1, \ldots, t_\xi)$ at any given point $(t_1, \ldots, t_\xi)$. To accomplish this, we can decompose the original system MDD into N_f sub-MDDs subMDD$_1, \ldots,$ subMDD$_{N_f}$ with following steps:

Step-1: Keep the terminal ξ and delete other terminal node;

Step-2: Tranverse upwards and delete all the nodes (except for the terminal nodes) that have no child node;

Step-3: Simplify the obtained MDD by merging isomorphic subtrees.

With the above three-step procedure, we can obtain the sub-MDD subMDD$_\xi$ that corresponds to the event that "the system experiences ξ covered component failures during the mission and the system survives the mission". Consider the example MDD in Fig. 4.7. It can be decomposed into two sub-MDD, as shown in Fig. 4.8, where every sub-MDD contains only the paths that lead to the same terminal value.

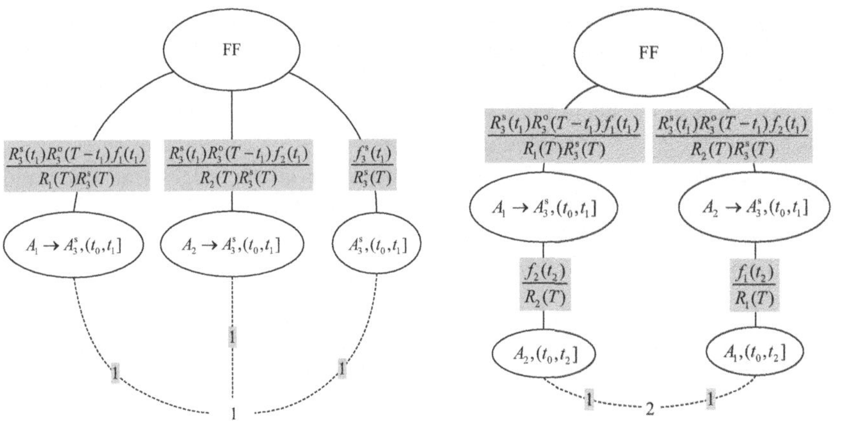

(a) The sub-MDD leads to the terminal 1, sub$\mathbf{MDD_1}$ (b) The sub-MDD leads to the terminal 2, sub$\mathbf{MDD_2}$

Fig. 4.8 Sub-MDDs for the example three-component system after decomposition according to the terminal values

4.5.2 Reassign of Edge Values in System MDD

After decomposition, each subMDD contains only the paths that lead to the same terminal value, and the occurrence probability associated with each subMDD corresponds to a single term in the system reliability in (4.13). For a subMDD$_\xi$, to computer this probability numerically, we need to evaluate the value of the integrand $G_\xi(t_1, \ldots, t_\xi)$. As mentioned, $G_\xi(t_1, \ldots, t_\xi)$ can be viewed as a probability density, and its can be obtained in a similar way as the system reliability in Sect. 4.4.2. To accomplish this, we should reassign "probability density", instead on "probability", to each edge in the sub-MDD. Corresponding to a general edge that lead to the node in (4.6), we assign the following probability density to this edge: for the

$$\frac{\prod_{v=1}^{r}\left[R_{j_v}^{\mathrm{s}}(t_{l_2})R_{j_v}^{\mathrm{o}}(T-t_{l_2})\right]f_i^*(t_{l_2}-t_{l_1})}{\prod_{v=1}^{r}R_{j_v}^{\mathrm{s}}(T)\times R_i^*(T-t_{l_1})}. \tag{4.14}$$

Note that this expression is just the integrand in (4.7). The probability density associated with each edge in the sub-MDD is listed in Table 4.1. As all the paths in a sub-MDD lead to a same terminal, we assign the edges leading to the terminal value with the probability density 1, and take the FLC effect into account only in the final calculation of the occurrence probability. The probability density assignment for each edge in Fig. 4.8 is marked in the figure.

After the assignment of probability densities for all the edges, the integrand $G_\xi(t_1, \ldots, t_\xi)$ can be obtained as the cumulative probability density associated with the topmost node in a bottom-up way as in Sect. 4.4.2. For example, $G_1(t_1)$ for the subMDD$_1$ in Fig. 4.8a is

$$G_1(t_1) = \frac{f_1(t_1)R_3^{\mathrm{s}}(t_1)R_3^{\mathrm{o}}(T-t_1)}{R_1(T)R_3^{\mathrm{s}}(T)} + \frac{f_2(t_1)R_3^{\mathrm{s}}(t_1)R_3^{\mathrm{o}}(T-t_1)}{R_2(T)R_3^{\mathrm{s}}(T)} + \frac{f_3^{\mathrm{s}}(t_1)}{R_3^{\mathrm{s}}(T)}, \tag{4.15}$$

Table 4.1 Probability density associated with each edge in the sub-MDD

Node type	Probability density associated with the edge leading to the node
$A_i \to \left(A_{j_1}^{\mathrm{s}}, \ldots, A_{j_r}^{\mathrm{s}}\right), (t_0, t_l]$	$\frac{\prod_{v=1}^{r}\left[R_{j_v}^{\mathrm{s}}(t_l)R_{j_v}^{\mathrm{o}}(T-t_l)\right]f_i(t_l)}{\prod_{v=1}^{r}R_{j_v}^{\mathrm{s}}(T)\times R_i(T)}$
$A_i, (t_0, t_l]$ or $A_i^{\mathrm{s}}, (t_0, t_l]$	$\frac{f_i(t_l)}{R_i(T)}$ 或 $\frac{f_i^{\mathrm{s}}(t_l)}{R_i^{\mathrm{s}}(T)}$
$A_i^{\mathrm{o}}, (t_{l_1}, t_{l_2}]$	$\frac{f_i^{\mathrm{o}}(t_{l_2}-t_{l_1})}{R_i^{\mathrm{o}}(T-t_{l_1})}$
$A_i^{\mathrm{o}} \to \left(A_{j_1}^{\mathrm{s}}, \ldots, A_{j_r}^{\mathrm{s}}\right), (t_{l_1}, t_{l_2}]$	$\frac{\prod_{v=1}^{r}\left[R_{j_v}^{\mathrm{s}}(t_{l_2})R_{j_v}^{\mathrm{o}}(T-t_{l_2})\right]f_i^{\mathrm{o}}(t_{l_2}-t_{l_1})}{\prod_{v=1}^{r}R_{j_v}^{\mathrm{s}}(T)\times R_i^{\mathrm{o}}(T-t_{l_1})}$
Ternimal node ξ	1

$G_2(t_1)$ for the subMDD$_2$ in Fig. 4.8b is

$$G_2(t_1, t_2) = \frac{f_1(t_1)R_3^{\mathrm{s}}(t_1)R_3^{\mathrm{o}}(T - t_1)}{R_1(T)R_3^{\mathrm{s}}(T)}\frac{f_2(t_2)}{R_2(T)} + \frac{f_2(t_1)R_3^{\mathrm{s}}(t_1)R_3^{\mathrm{o}}(T - t_1)}{R_2(T)R_3^{\mathrm{s}}(T)}\frac{f_1(t_2)}{R_1(T)}. \tag{4.16}$$

4.5.3 Numerical Calculation of Occurrence Probabilities

Consider a sub-MDD subMDD$_\xi$. It contains all the possible scenarios that the system experiences ξ component failures during the mission time T and still succeeds the mission. The occurrence probability of this event is:

$$\Pr\left(T|\text{CMDD}_\xi\right) = R_0(T) \times \prod_{j=1}^{\xi} c_j \times \int_{\Delta_T^\xi} G_\xi\left(t_1, \ldots, t_\xi\right)dt_1 \ldots dt_\xi. \tag{4.17}$$

Because the integrand is quite complicate in general, we have to resort to numerical methods to obtain this integral. The numerical integration approximates the integral by a weighted sum of the values of the integrand on selected points:

$$\int_{\Delta_T^\xi} G_\xi\left(t_1, \ldots, t_\xi\right)dt_1 \ldots dt_\xi \approx \alpha\left(\Delta_T^\xi\right)\sum_{j=1}^{m_0} w_j G_\xi\left(t_j^\xi\right), \tag{4.18}$$

where $\alpha\left(\Delta_T^\xi\right)$ is a constant related to the size of the integration region, $w_j, j = 1, \ldots, m_0$ are the weight, and $t_j^\xi = \left(t_{1,j}, \ldots, t_{\xi,j}\right)$ are the selected points.

In our problem, the integration region $\Delta_T^\xi = \left\{\left(t_1, \ldots, t_\xi\right)|0 < t_1 < \ldots < t_\xi < T\right\}$ is a ξ-simplex. According to [1], the numerical integration can be implemented as follows.

Step-1: Set the accuracy criterion ε_0 and maximum number of sampling points M. Let $f_{eva} = 0$.

Step-2: Let $S_{tmp} = \Delta_T^{\xi}$. Go to Step-7 and Step-8 to obtain the value of the integral and the error estimate Q_{tmp} and ε_{tmp}, respectively. If $\varepsilon_{tmp} < \varepsilon_0$, then the accuracy of the integration is acceptable, and return Q_k^0 and finish. Otherwise, go to Step-3.

Step-3: Let $\boldsymbol{S} = \left\{\Delta_T^{\xi}\right\}$, $\boldsymbol{Q} = \left\{Q_{tmp}\right\}$, $\boldsymbol{E} = \left\{\varepsilon_{tmp}\right\}$, $Q_S = Q_{tmp}$, $E_S = \varepsilon_{tmp}$.

Step-4: Find the maximum ε_l in $\boldsymbol{E}$ and the corresponding element Q_l in $\boldsymbol{Q}$ and S_l in $\boldsymbol{S}$. Cut S_l into two simplexes S_{l1} and S_{l2}, and go to Step-7 and Step-8 to calculate the corresponding integrals and the error estimates $Q_{tmp}^1, \varepsilon_{tmp}^1$, $Q_{tmp}^2, \varepsilon_{tmp}^2$.

Step-5: Let $\boldsymbol{S} = (\boldsymbol{S}\backslash\{S_l\}) \cup \{S_{l1}, S_{l2}\}$, $\boldsymbol{Q} = (\boldsymbol{Q}\backslash\{Q_l\}) \cup \left\{Q_{tmp}^1, Q_{tmp}^2\right\}$, $\boldsymbol{E} = (\boldsymbol{E}\backslash\{\varepsilon_l\}) \cup \left\{\varepsilon_{tmp}^1, \varepsilon_{tmp}^2\right\}$, $Q_S = Q_S - Q_l + Q_{tmp}^1 + Q_{tmp}^2$, $E_S = E_S - \varepsilon_{tmp} + \varepsilon_{tmp}^1 + \varepsilon_{tmp}^2$;

Step-6: If $E_S \le \varepsilon_0$, then return Q_S and finish. Otherwise, go to Step-4.

Step-7: Sample m_0 points $\boldsymbol{t}_1^{\xi}, \ldots, \boldsymbol{t}_{m_0}^{\xi}$ inside the simplex S_{tmp}, calculate each $G_{\xi}\left(\boldsymbol{t}_j^{\xi}\right)$, $j = 1, \ldots, m_0$, and obtain the integral Q_{tmp} on S_{tmp} and the error estimate ε_{tmp}.

Step-8: $f_{eva} = f_{eva} + m_0$. If $f_{eva} > M$, then return the current Q_S and E_S and terminate the calculation.

In this algorithm, the simplex cutting method in Step-4, and the way to select points and calculate Q_{tmp} and ε_{tmp} in Step-7 can be found in [1, 2]. Note that the value of $G_{\xi}\left(\boldsymbol{t}_j^{\xi}\right)$ at a selected point $\boldsymbol{t}_j^{\xi} = \left(t_{1,j}, \ldots, t_{\xi,j}\right)$. can be easily calculated following the bottom-up manner in Sect. 4.5.2.

4.5.4 Programs for Autatically System Reliability Calculation

Using the numerical integration method, the occurence probability associated with every sub-MDD can be obtained. Then, the system reliability can be obtained as

$$R_S(T) = R_0(T) + \sum_{\xi=1}^{N_f} \Pr\left(T|\text{CMDD}_{\xi}\right). \tag{4.19}$$

Furthermore, other reliability indices, such as the expected lifetime, can be obtained numerically. For example, the expected lifetime of the system can be obtained using the Gaussian-Laguerre method:

$$MTTF = \int_0^{+\infty} R_S(t)dt \approx \sum_{j=1}^{m} w_j e^x R_S\left(t_j\right).$$

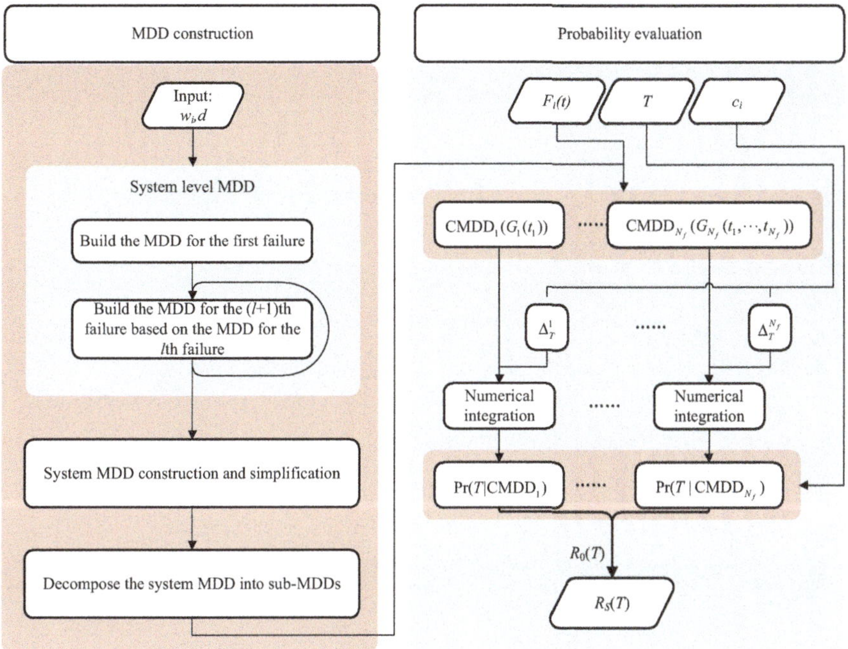

Fig. 4.9 Procedure to evaluation the system reliability based on the MDD method

To summarize, the system reliability evaluation of the warm standby system involves two main parts, i.e.,the system MDD construction and the numerical calculation of the occurrence probabilities. Figure 4.9 illustrates the flow to calculate the system reliability based on the MDD method.

References

1. Genz A, Cools R. An adaptive numerical cubature algorithm for simplices [J]. ACM Transactions on Mathematical Software (TOMS), 2003, 29(3): 297–308.
2. Berntsen J, Espelid T O. Error estimation in automatic quadrature routines [J]. ACM Transactions on Mathematical Software (TOMS), 1991, 17(2): 233–52.

Chapter 5
Reliability of Warm Standby Systems with Imperfect Fault Coverage and Switching Failure

The redundant system with warm standby components is still considered in this chapter. Different from Chap. 3, this chapter assumes that the fault coverage mechanism of the copmonent is element level fault coverage, not fault level fault coverage. In addition, we also consider the effect of switching failure. The method is still based on multi-valued decision diagram [1–4]. We extend it to the case where we can study the sequence dependency, so as to model the redundant system with warm standby components considering element level fault coverage. This extended method is called sequence MDD (SMDD) method. This method is suitable for the reliability calculation of warm standby redundant system with arbitrary failure time distribution, and the influence of fault coverage and switching failure can be considered.

5.1 Reliability Model Under Perfect Switching

This section describes the proposed method for evaluating the reliability of a WSP system with components subject to imperfect fault coverage. The traditional MDD method has been used for the reliability evaluation of a static multi-state system. To consider the time-dependent dynamic behavior of warm spares, the traditional MDD construction and evaluation rules need to be modified. The proposed method can be described as a three-step procedure based on SMDD: (1) Variable encoding and dynamic fault tree (DFT) conversion; (2) Generating the SMDD; and (3) Evaluating the SMDD. These three steps are elaborated in the following subsections.

5.1.1 Variable Encoding and DFT Conversion

Warm standby sparing behavior has been modeled using a WSP gate in the DFT analysis [1, 4]. Figure 5.1 illustrates the general structure of a WSP gate with n spares,

R. Peng et al., *Reliability Modelling and Optimization of Warm Standby Systems*,
https://doi.org/10.1007/978-981-16-1792-8_5

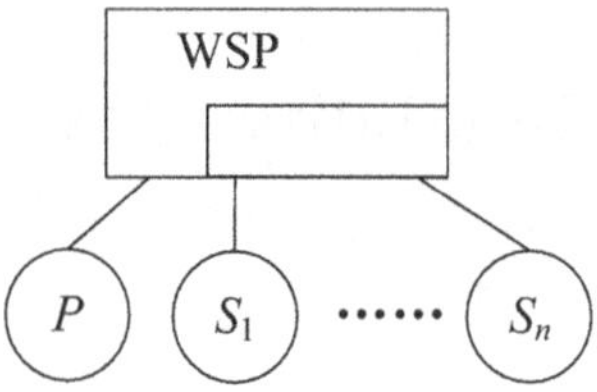

Fig. 5.1 WSP Gate with n-spares

where P represents the primary component and S_i represents a warm-standby unit. All the spares are used in the specified order (from left to right) as long as it is still operational.

The gate fires when any component fails globally or when all components fail locally. Therefore, the WSP gate can be replaced with an OR gate. All the components failing globally in addition to an AND gate with all components failing locally are the inputs of the OR gate. The WSP gate is then converted to the fault tree model shown in Fig. 5.2. In the case of DFT with multiple non-independent WSP gates, sequential basic events are introduced into the DFT conversion. Such a case is illustrated in the case study. The SMDD encoding of the primary and warm standby components is described next.

Each primary component in the proposed method is associated with a triple-valued variable represented by a SMDD node with three outgoing edges as shown in Fig. 5.3. The three edges represent the three states of the primary component: operational (0), failed locally (L), and failed globally due to the IPC (G), respectively.

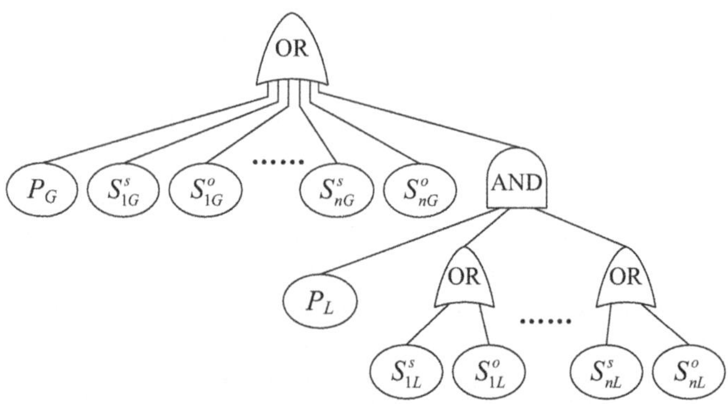

Fig. 5.2 The converted DFT model

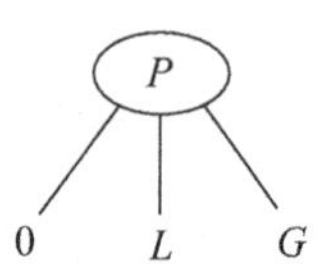

Fig. 5.3 Representation of primary component in SMDD

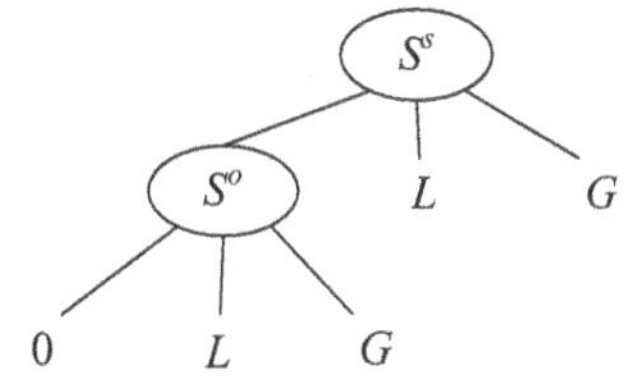

Fig. 5.4 Representation of spare component in SMDD

Figure 5.4 illustrates the SMDD model of a warm standby component. The time dependence specifically the failure rate of a warm standby component changing from a reduced rate α to a full failure rate λ after replacing the faulty primary component is taken into consideration by using two nodes S^s and S^o.

5.1.2 SMDD Construction

The second step is to construct the SMDD. The basic MDD manipulation rules are extended with the following rules to incorporate the consideration of the time-dependent failure behavior of warm standby components.

Rule 1: For a primary component P, $(P|_G \text{ op } Y) = 1$; meaning that in the SMDD construction if the primary unit is in the global failure state, regardless of the type of the component Y and the operation applied, the resultant node is the sink node '1′ (the entire system failure);

Rule 2: For the spare component S, if the primary component does not fail (valued 0), then $(P|_0 \text{ op } S(\alpha)|_G) = 1$, meaning that if the spare fails globally before the primary unit fails, the entire system fails;

Rule 3: $(\mathrm{P}|_0 \text{ op } S(\lambda)|_G) = 0$ and $(S_i(\lambda)|_0 \text{ op } S_j(\lambda)|_G) = 0$ $(i < j)$ meaning that if the primary unit does not fail, the spare cannot fail with a full failure rate.

Note that $X|_L$ and $X|_G$ (representing component X being in the local failure state L and the global failure state G, respectively) correspond to "1" in the final SMDD model.

5.1.3 System Reliability Calculation

The last step is to evaluate the system SMDD to obtain the final system reliability/unreliability. Similar to the traditional MDD evaluation, in the SMDD model the unreliability of a WSP system can be evaluated as the sum of probabilities of all the disjoint paths from the root node to sink node "1". In this step, the dependence among all components is handled. In particular, the sequential failure of the warm standby component is considered for both cases: the standby unit fails with a reduced failure rate before the primary component fails ($S_L^s \rightarrow P_L$), and the standby unit fails with a full failure rate when it fails after replacing the faulty primary component ($P_L \to S_L^o$). These sequence/temporal dependencies must be considered in the probability evaluation. The probability that the spare fails before the primary component is given by

$$\Pr\{S_L^s \to P_L\} = \int_0^t \int_0^{\tau_1} g_{P_L}^o(\tau_1) g_{S_L}^s(\tau_2) d\tau_2 d\tau_1. \tag{5.1}$$

where $g_{P_L}^o(\tau_1) = f_{P_L}^o(\tau_1) R_{P_G}^o(\tau_1)$ is the probability density function (pdf) of time-to-local-failure of the primary component when it fails locally but not globally; similarly $g_{S_L}^s(\tau_2) = f_{S_L}^s(\tau_2) R_{S_G}^s(\tau_2)$ is the pdf of time-to-local-failure of the spare component when it fails locally but not globally in the standby mode.

The failure distribution of the standby component after it is put into operation generally depends on the switch-over-time. The accelerated life testing (ALT) is a process to evaluate the impact of stress on a product or component [5, 6]. Thus, ALT is used for studying time-dependent reliability characteristic of load-sharing systems by extrapolating the results of the statistical analysis [8]. The accelerated failure time model (AFTM) assumes that the effect of the load on the lifetime of a component is multiplicative in time, that is, $F_{sb}(t) = F_{op}(t.\varphi_s)$ where $F_{sb}(t)$, $F_{op}(t)$ and φ_s denote the failure probability in standby mode, the failure probability operational mode, and the acceleration (deceleration) rate, respectively. The proportional hazards model (PHM) coincides with AFTM only when the baseline reliability distribution is Weibull [10]. PHM assumes the effect of the covariate is multiplicative on the failure rate: $F_{sb}(t) =$ 1-(1-$F_{op}(t))^{\varphi sb}$, where φ_{sb} is the multiplicative factor. Using the AFTM it is easy to interpret the results and the regression parameters estimates are robust to the omitted covariate. There are three types of ALT: in a *constant-stress ALT*, the stress applied to the items is constant during the test; a *step-stress ALT* changes the stress at specific times; and a *progressive-stress ALT* increases the stress continuously. The impact of the load history on the component failure rate is considered, since the component operates at different loads during different time intervals. For the step-stress ALT there are mainly three models that consider the load history effect on a component failure rate: tampered random variable (TRV) [7], cumulative exposure model (CEM) [8, 11], and tampered failure rate (TFR) [5, 9]. The CEM takes into consideration the cumulative effect of the applied stresses in the reliability calculation. The cumulative failure rate is calculated using a weighted sum of all load durations. For the TFR the change in the stress level is reflected in the hazard rate. The failure rate only depends on the current load and the age of the component and the reliability is function of the loads history and age at which the loads have been applied. This work applies the AFTM with the concept of CEM. Thus the probability that the spare fails after the primary unit with a full failure rate λ is given by:

$$\begin{aligned}\Pr\{P_L \to S_L^o\} &= \int_0^t \int_{\tau_1}^t g_{P_L}^o(\tau_1) g_{S_L}^o(\tau_2 - \tau_1 + \gamma\tau_1)\left(1 - \int_0^{\tau_1} f_S(\tau)d\tau\right)d\tau_2 d\tau_1 \\ &= \int_0^t \int_{\tau_1}^t g_{P_L}^o(\tau_1) g_{S_L}^o(\tau_2 - \tau_1 + \gamma\tau_1) R_S^s(\tau_1) d\tau_2 d\tau_1.\end{aligned} \tag{5.2}$$

where, $g^o_{S_L}(\tau_2-\tau_1+\gamma\tau_1) = f^o_{S_L}(\tau_2-\tau_1+\gamma\tau_1)\left(1-\int_{\tau_1}^{\tau_2} f^o_{S_G}(\tau_3-\tau_1+\gamma\tau_1)d\tau_3\right)$ 。The "S" in subscript of $R^s_S(\tau_1)$ does not include "L"or "G",indicating the reliability considering both local and global failure, that is, $R^s_S(\tau_1) = R^s_{S_L}(\tau_1)R^s_{S_G}(\tau_1)$.

The *pdf* in (5.1) and (5.2) can follow any distribution. In the case of an exponential distribution, (5.1) and (5.2) can be respectively rewritten as

$$\begin{aligned}\Pr\{S^s_L \to P_L\} &= \int_0^t\int_0^{\tau_1} \lambda_{P_L} e^{-\lambda_{P_L}\tau_1} e^{-\lambda_{P_G}\tau_1} \alpha_{S_L} e^{-\alpha_{S_L}\tau_2} e^{-\alpha_{S_G}\tau_2} d\tau_2 d\tau_1 \\ &= \int_0^t\int_0^{\tau_1} \lambda_{P_L} e^{-\lambda_P\tau_1} \alpha_{S_L} e^{-\alpha_S\tau_2} d\tau_2 d\tau_1\end{aligned} \tag{5.3}$$

$$\begin{aligned}\Pr\{P_L \to S^o_L\} &= \int_0^t\int_{\tau_1}^{t} \lambda_{P_L} e^{-\lambda_{P_L}\tau_1} e^{-\lambda_{P_G}\tau_1} \lambda_{S_L} e^{-\lambda_{S_L}(\tau_2-\tau_1)} \\ &\quad \cdot \left(1-\int_{\tau_1}^{\tau_2} \lambda_{S_G} e^{-\lambda_{S_G}(\tau-\tau_1)} d\tau\right) e^{-\alpha_S\tau_1} d\tau_2 d\tau_1 \\ &= \int_0^t\int_{\tau_1}^{t} \lambda_{P_L} e^{-\lambda_P\tau_1} \lambda_{S_L} e^{-\lambda_S(\tau_2-\tau_1)} e^{-\alpha_S\tau_2} d\tau_2 d\tau_1.\end{aligned} \tag{5.4}$$

where λ_{S_L} and λ_{S_G} are the spare component's local and global full failure rate, respectively; $\lambda_S = \lambda_{S_L} + \lambda_{S_G}$. α_{S_L} and α_{S_G} are the spare component's local and global reduced failure rate, respectively;$\alpha_S = \alpha_{S_L} + \alpha_{S_G}$. λ_{P_L} and λ_{P_G} are the primary component's local and global full failure rate, respectively; $\lambda_P = \lambda_{P_L} + \lambda_{P_G}$.

Figure 5.5 illustrates the DFT model of a warm standby sparing sensor system, where A is the primary sensor node, and S_1 and S_2 are the standby sensors. The spare sensor i ($i = 1, 2$) fails with a reduced failure rate α_{si} before the primary sensor fails, and fails with a regular failure rate λ_{si} after the primary fails, where $\lambda_{si} > \alpha_{si}$. The reliability of this example WSP system is evaluated using the proposed SMDD method as follows.

First, the example WSP system is converted into a DFT model as shown in Fig. 5.6.

Figure 5.7 shows the system SMDD construction by applying the logic AND operation between SMDD of the primary component A and the SMDD of the spare components S_1 and S_2. Figure 5.8 shows the final SMDD after removing the invalid nodes and related branches.

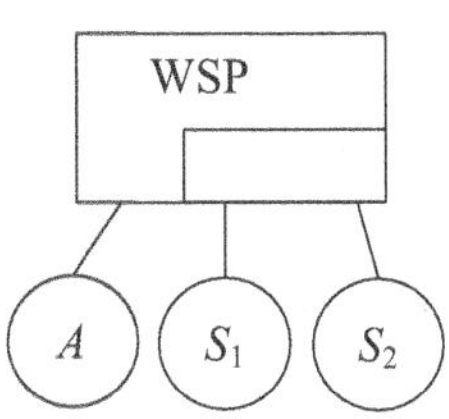

Fig. 5.5 An illustrative WSP example

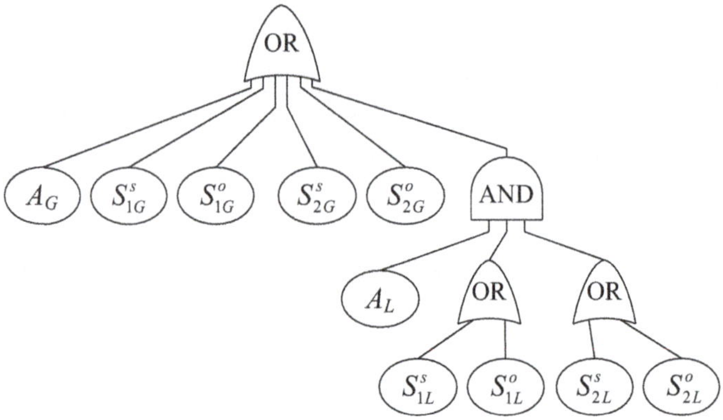

Fig. 5.6 The converted DFT model

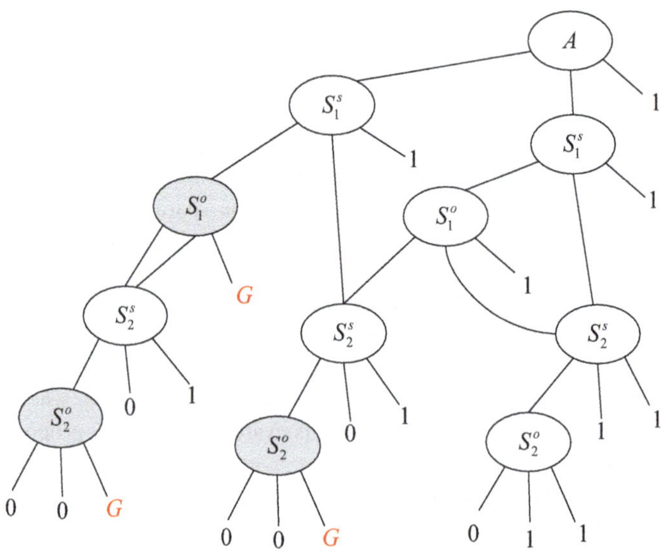

Fig. 5.7 SMDD of the example WSP

The system unreliability is the sum of probabilities of all the paths from the root to sink node "1" (system failure). For the SMDD of this example in Fig. 5.8, there are 15 paths that lead to the sink node "1":

1:A_G;	2: $A_L \sim S^s_{1G}$;	3:$A_L \sim S^s_{1L} \sim S^s_{2G}$;
4: $A_L \sim S^s_{1L} \sim S^s_{2L}$;	5:$A_L \sim S^s_{1L} \sim S^o_{2G}$;	6: $A_L \sim S^s_{1L} \sim S^o_{2L}$;
7:$A_L \sim S^o_{1L} \sim S^s_{2L}$;	8: $A_L \sim S^o_{1L} \sim S^s_{2G}$;	9:$A_L \sim S^o_{1L} \sim S^o_{2L}$;
10: $A_L \sim S^o_{1L} \sim S^o_{2G}$;	11: $A_L \sim S^o_{1G}$;	12: $A_L \sim S_1 \sim S^s_{2G}$;

(continued)

(continued)

1:A_G;	2: $A_L \sim S_{1G}^s$;	3:$A_L \sim S_{1L}^s \sim S_{2G}^s$;
13: $A \sim S_{1G}^s$;	14:$A \sim S_{1L}^s \sim S_{2G}^s$;	15: $A \sim S_1^s \sim S_{2G}^s$

The symbol '$\sim$' denotes the ordering of the events in the path starting from the root node in the SMDD. Assume ($S^s = S_L^s + S_G^s$). The paths 3 and 4 can be merged into $A_L \sim S_{1L}^s \sim S_2^s$; paths 7 and 8 are merged into $A_L \sim S_{1L}^o \sim S_2^o$. Similarly, paths 5 and 6 are merged into one path $A_L \sim S_{1L}^s \sim S_2^o$ and paths 9 and 10 can be merged into $A_L \sim S_{1L}^o \sim S_2^o$ by assuming $A_L \sim S_{1L}^o \sim S_2^o$. Lastly, paths 14 and 15 can be merged to $A \sim S_{1G}^s \sim S_{2G}^s$. After the reduction the number of paths is reduced to 10. In the probability analysis the sequential order is defined by the standby component's failure mode. If the component fails with a reduce failure rate, it means that it fails before the primary component. For example, $A_L \sim S_{1G}^s$ corresponds to the sequential event $S_{1G}^s \to A_L$. Otherwise, if the standby fails with a full failure rate, e.g., $A_L \sim S_{1G}^o$ the sequential event is given by $S_{1G}^o \to A_L$. The system unreliability is then computed as:

$$\begin{aligned} F_S(t) = & \Pr\{A_G\} + \Pr\{S_{1G}^s \to A_L\} + \Pr\{(S_{1L}^s \wedge S_2^s) \to A_L\} + \Pr\{S_{1L}^s \to A_L \to S_2^o\} \\ & + \Pr\{A_L \to S_{1L}^o \wedge S_2^s\} + \Pr\{A_L \to S_{1L}^o \to S_2^o\} \\ & + \Pr\{A_L \to S_{1G}^o\} + \Pr\{S_{2G}^s \wedge A_L\} \Pr\{\neg S_1\} \\ & + \Pr\{\neg A \wedge S_{1G}^s\} + \Pr\{\neg A\} \Pr\{\neg S_1^s\} \Pr\{S_{2G}^s\} \end{aligned} \tag{5.5}$$

where, "$\wedge$" denotes logic "AND". Let $T_A, T_{S_{iL}}^s, T_{S_{iG}}^s$ and $T_{S_i}^o$ be failure time of A, S_{iL}^s, S_{iG}^s and S_i^o, and $f_A(t)$, $f_{S_{iL}}^s(t)$, $f_{S_{iG}}^s$ and $f_{S_i}^o(t)$ be their PDF. According to (5.1) and (5.2), it obtains:

$$\Pr\{S_{1G}^s \to A_L\} = \int_0^t \int_0^{\tau_1} g_{A_L}^o(\tau_1) f_{S_{1G}}^s(\tau_2) d\tau_2 d\tau_1,$$

$$\Pr\{(S_{1L}^s \wedge S_2^s) \to A_L\} = \int_0^t \int_0^{\tau_1} \int_0^{\tau_1} g_{A_L}^o(\tau_1) g_{S_{1L}}^s(\tau_2) f_{S_2}^s(\tau_3) d\tau_3 d\tau_2 d\tau_1$$

$$\begin{aligned} & \Pr\{S_{1L}^s \to A_L \to S_2^o\} \\ = & \int_0^t \int_0^{\tau_1} \int_{\tau_1}^t g_{A_L}^o(\tau_1) g_{S_{1L}}^s(\tau_2) f_{S_2}^o(\tau_3 - \tau_1 + \gamma_{S_1}\tau_1) \left(1 - \int_0^{\tau_1} f_{S_2}^s(\tau) d\tau\right) d\tau_3 d\tau_2 d\tau_1. \end{aligned} \tag{5.6}$$

Similarly, we have

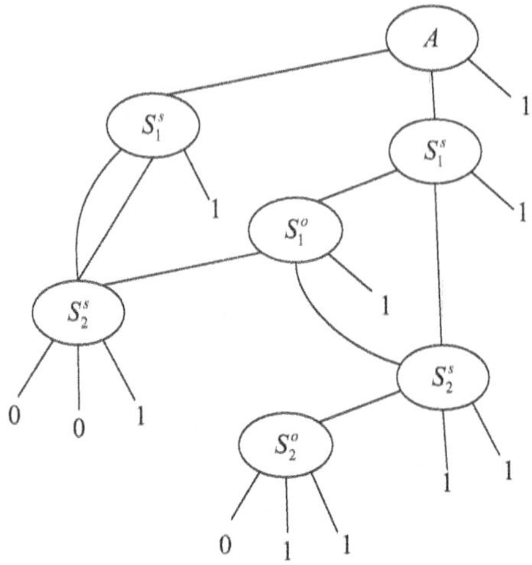

Fig. 5.8 The final SMDD of a WSP with two spares

$$
\begin{aligned}
&\Pr\{A_L \to S_{1L}^o \wedge S_2^s\} \\
&= \int_0^t \int_{\tau_1}^t \int_0^{\tau_2} g_{A_L}^o(\tau_1) g_{S_1}^o(\tau_2 - \tau_1 + \gamma_{S_1}\tau_1)\left(1 - \int_0^{\tau_1} f_{S_1}^s(\tau)d\tau\right) f_{S_2}^s(\tau_3)d\tau_3 d\tau_2 d\tau_1, \\
&\Pr\{A_L \to S_{1L}^o \to S_2^o\} \\
&= \int_0^t \int_{\tau_1}^t \int_{\tau_2}^t g_{A_L}^o(\tau_1) g_{S_{1L}}^o(\tau_2 - \tau_1 + \gamma_{S_1}\tau_1)\left(1 - \int_0^{\tau_1} f_{S_1}^s(\tau)d\tau\right) \\
&\cdot f_{S_2}^o(\tau_3 - \tau_2 + \gamma_{S_1}\tau_2)\left(1 - \int_0^{\tau_2} f_{S_2}^s(\tau)d\tau\right) d\tau_3 d\tau_2 d\tau_1, \\
&\Pr\{A_L \to S_{1G}^o\} = \int_0^t g_{A_L}^o(\tau_1) \int_{\tau_1}^t f_{S_{1G}}^o(\tau_2 - \tau_1 + \gamma_{S_1}\tau_1)\left(1 - \int_0^{\tau_1} f_{S_1}^s(\tau)d\tau\right) d\tau_2 d\tau_1, \\
&\Pr\{S_{2G}^s \wedge A_L\}\Pr\{\neg S_1\} = \int_0^t \int_0^t g_{A_L}^o(\tau_1)d\tau_1 f_{S_{2G}}^s(\tau_3)d\tau_3 \\
&\cdot\left(1 - \int_0^{\tau_1} f_{S_1}^s(\tau_2)d\tau_2\right)\left(1 - \int_{\tau_1}^t f_{S_1}^o(\tau_2 - \tau_1 + \gamma_{S_1}\tau_1)d\tau_2\right) \\
&\Pr\{\neg A \wedge S_{1G}^s\} = \int_0^t f_{S_{1G}}^s(\tau_2)d\tau_2\left(1 - \int_0^t f_A^o(\tau_1)d\tau_1\right), \\
&\Pr\{\neg A \wedge S_{2G}^s\}\Pr\{\neg S_1^s\} = \int_0^t f_{S_{2G}}^s(\tau_3)d\tau_3\left(1 - \int_0^t f_A^o(\tau_1)d\tau_1\right)\left(1 - \int_0^t f_{S_1}^s(\tau_2)d\tau_2\right).
\end{aligned}
\tag{5.7}
$$

Using the results of (5.5), (5.6) and (5.7), the system unreliability can be written as:

$$
\begin{aligned}
F_S(t) &= \int_0^t f_{A_G}^o(\tau_1)d\tau_1 \\
&+ \int_0^t \int_{\tau_1}^t \int_{\tau_2}^t g_{A_L}^o(\tau_1) g_{S_{1L}}^o(\tau_2 - \tau_1 + \gamma_{S_1}\tau_1)\left(1 - \int_0^{\tau_1} f_{S_1}^s(\tau)d\tau\right) \\
&\cdot f_{S_2}^o(\tau_3 - \tau_2 + \gamma_{S_2}\tau_2)\left(1 - \int_0^{\tau_2} f_{S_2}^s(\tau)d\tau\right)d\tau_3 d\tau_2 d\tau_1 \\
&+ \int_0^t \int_0^{\tau_1} \int_0^{\tau_1} g_{A_L}^o(\tau_1) g_{S_{1L}}^s(\tau_2) g_{S_2}^s(\tau_3) d\tau_3 d\tau_2 d\tau_1 \\
&+ \int_0^t \int_0^{\tau_1} \int_{\tau_1}^t g_{A_L}^o(\tau_1) g_{S_{1L}}^s(\tau_2) f_{S_2}^o(\tau_3 - \tau_1 + \gamma_{S_2}\tau_1)\left(1 - \int_0^{\tau_1} f_{S_2}^s(\tau)d\tau\right)d\tau_3 d\tau_2 d\tau_1 \\
&+ \int_0^t \int_{\tau_1}^t \int_0^{\tau_2} g_{A_L}^o(\tau_1) f_{S_1}^o(\tau_2 - \tau_1 + \gamma_{S_1}\tau_1)\left(1 - \int_0^{\tau_1} f_{S_1}^s(\tau)d\tau\right) f_{S_2}^s(\tau_3)d\tau_3 d\tau_2 d\tau_1 \\
&+ \int_0^t f_{S_{1G}}^s(\tau_2)d\tau_2\left(1 - \int_0^t f_A^o(\tau_1)d\tau_1\right) \\
&+ \int_0^t g_{A_L}^o(\tau_1) \int_{\tau_1}^t f_{S_{1G}}^o(\tau_2 - \tau_1 + \gamma_{S_1}\tau_1)\left(1 - \int_0^{\tau_1} f_{S_1}^s(\tau)d\tau\right)d\tau_2 d\tau_1 \\
&+ \int_0^t f_{S_{2G}}^s(\tau_3)d\tau_3\left(1 - \int_0^t f_A^o(\tau_1)d\tau_1\right)\left(1 - \int_0^t f_{S_1}^s(\tau_2)d\tau_2\right) \\
&+ \int_0^t \int_0^t g_{A_L}^o(\tau_1)d\tau_1 f_{S_{2G}}^s(\tau_3)d\tau_3\left(1 - \int_0^{\tau_1} f_{S_1}^s(\tau_2)d\tau_2\right)\left(1 - \int_{\tau_1}^t f_{S_1}^o(\tau_2 - \tau_1 + \gamma_{S_1}\tau_1)d\tau_2\right) \\
&+ \int_0^t g_{A_L}^o(\tau_1) \int_0^{\tau_1} f_{S_{1G}}^s(\tau_2)d\tau_2 d\tau_1
\end{aligned}
\tag{5.8}
$$

We verify the proposed method by performing Markov analysis of the example system and comparing the results. Figure 5.9 illustrates the Markov model of the example system.

Table 5.1 summarizes the analysis results using both the proposed method and the Markov method. The following parameter values for the primary component and standby spares are used: $\lambda_{AL} = 0.36$/day, $\lambda_{AG} = 0.04$/day, $\alpha_{S1G} = 0.02$/day, $\alpha_{S1L} = 0.18$/day, $\lambda_{S1G} = 0.03$/day, $\lambda_{S1L} = 0.27$/day, $\alpha_{S2G} = 0.01$/day, $\alpha_{S2L} = 0.09$/day, $\lambda_{S2G} = 0.03$/day, $\lambda_{S1L} = 0.27$/day. The results using the two methods exactly match.

5.2 Case Study

5.2.1 Case Study One

Figure 5.10 illustrates the DFT model of a hard drive system, where *A*, *B* are the primary hard disks sharing the same warm spare disk *S*. The two primary hard disks

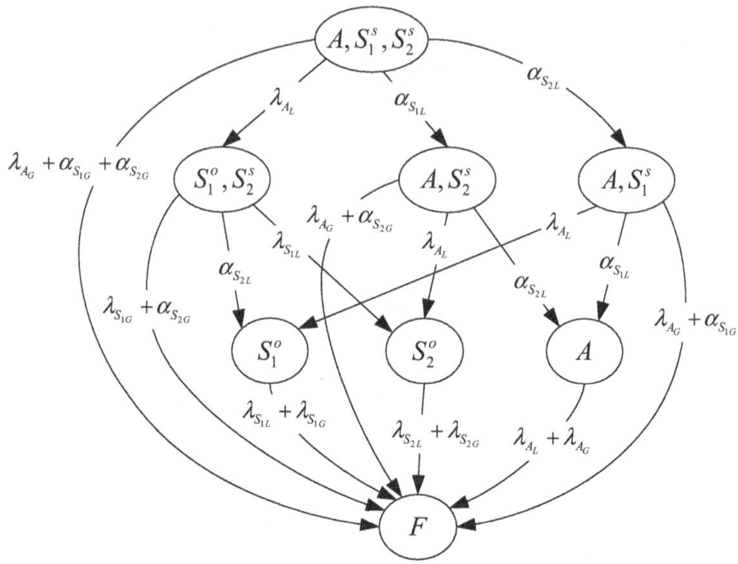

Fig. 5.9 Markov model of a WSP with two spares

Table 5.1 Results of both Markov and SMDD methods

Time	Markov	Proposed Method
5	0.4781	0.4781
10	0.8386	0.8386
15	0.9587	0.9587

Fig. 5.10 The DFT of a hard drive system

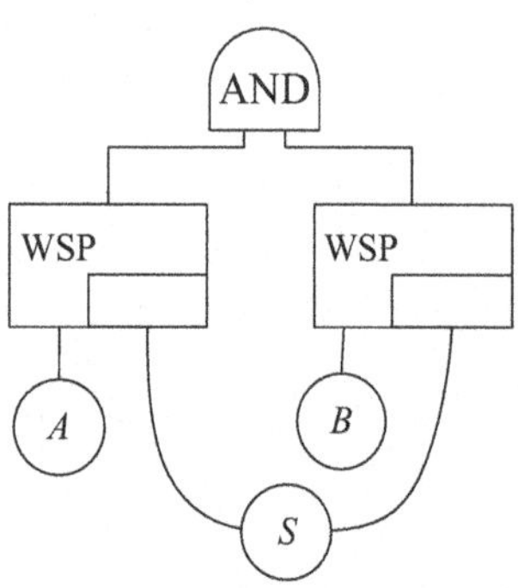

are working in parallel. The system fails when both WSP gates fire. The disk S fails with a reduced failure rate α_s any time before the failure of any of the primary components. Otherwise if any of the primary disks A or B fails first, the spare disk S replaces it and then fails with a full failure rate λ_s. The DFT in Fig. 5.10 is converted to an OR gate that connects any component failing globally, and an AND gate with the

Fig. 5.11 The converted DFT of the hard disk system

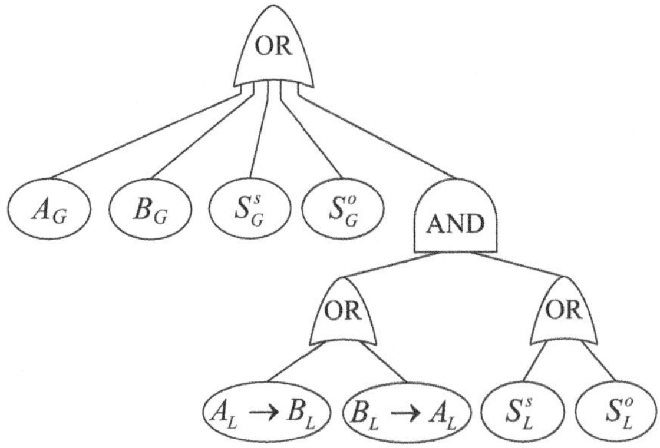

all component local failures as basic events. The two primary components share the same spare that replaces the first primary component to fail; therefore, it is important to consider the order in which the primary components fail and the DFT includes the two sequential events $A_L \to B_L$, $B_L \to A_L$. This is illustrated in Fig. 5.11 using the OR gate having two sequential events.

Figure 5.12 shows the final system SMDD model.

There are 12 paths in Fig. 5.12 from the root to the sink node "1" after the merging:

1: A_G;	2: $A_L \sim B_G$;	3: $A_L \to B_L \sim S^s$;
4: $A_L \to B_L \sim S^o$;	5: $B_L \to A_L \sim S^s$;	6: $B_L \to A_L \sim S^o$;
7: $A_L \sim B \sim S_G^s$;	8: $A_L \sim B \sim S_G^o$;	9: $A \sim B_G$;
10: $A \sim B_L \sim S_G^s$;	11: $A \sim B_L \sim S_G^o$;	12: $A \sim B \sim S_G^s$

The paths 2 and 9 are further merged into one path:: $A_G \sim B_G$. The probability of the sequential node is given as:

$$\Pr\{A_L \to B_L \wedge S_1^s\} = \int_0^t \int_{\tau_A}^t \int_0^{\tau_A} g_{A_L}(\tau_A) g_{B_L}(\tau_B) f_S^s(\tau_S) d\tau_S d\tau_B d\tau_A,$$

$$\Pr\{B_L \to A_L \wedge S_1^s\} = \int_0^t \int_{\tau_B}^t \int_0^{\tau_B} g_{B_L}(\tau_B) g_{A_L}(\tau_A) f_S^s(\tau_S) d\tau_s d\tau_A d\tau_B. \tag{5.9}$$

The system unreliability is the sum of probabilities of all the 10 paths as

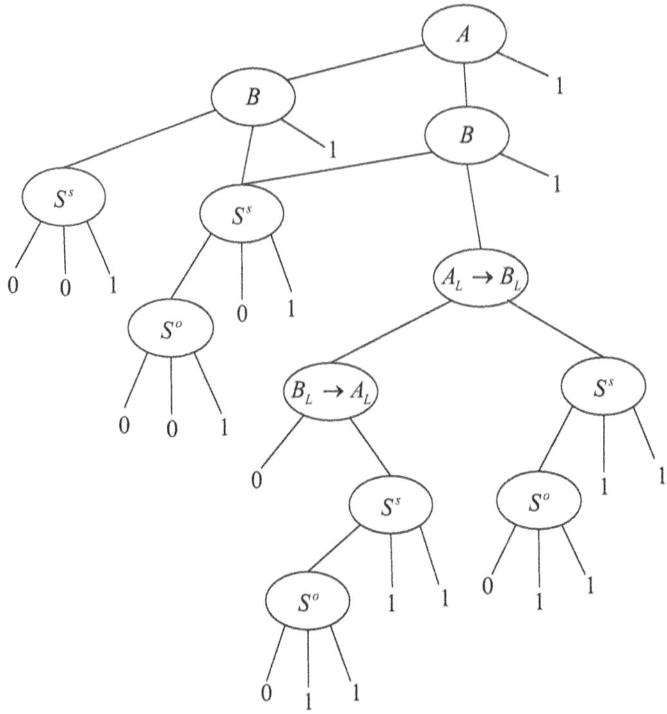

Fig. 5.12 Final SMDD of the disk example

$$
\begin{aligned}
F_S(t) &= \int_0^t f_{A_G}^o(\tau_A)d\tau_A + \left(1 - \int_0^t f_{A_G}^o(\tau_A)d\tau_A\right)\int_0^t f_{B_G}^o(\tau_B)d\tau_B \\
&+ \int_0^t g_{A_L}^o(\tau_A) \int_{\tau_A}^t g_{B_L}^o(\tau_B) \\
&\left[\int_0^{\tau_A} f_S^s(\tau_S)d\tau_S + \left(1 - \int_0^{\tau_A} f_S^s(\tau_S)d\tau_S\right)\int_{\tau_A}^t f_S^o(\tau_S - \tau_A + \gamma\tau_S)d\tau_S\right]d\tau_B d\tau_A \\
&+ \int_0^t g_{B_L}^o(\tau_B) \int_{\tau_B}^t g_{A_L}^o(\tau_A) \\
&\left[\int_0^{\tau_B} f_S^s(\tau_S)d\tau_S + \left(1 - \int_0^{\tau_B} f_S^s(\tau_S)d\tau_S\right)\int_{\tau_B}^t f_S^o(\tau_S - \tau_B + \gamma\tau_S)d\tau_S\right]d\tau_A d\tau_B \\
&+ \int_0^t g_{A_L}^o(\tau_A) \\
&\left(\int_0^{\tau_A} f_{S_G}^s(\tau_S)d\tau_S + \left(1 - \int_0^{\tau_A} f_S^s(\tau_S)d\tau_S\right)\cdot\int_{\tau_A}^t f_{S_G}^o(\tau_S - \tau_A + \gamma\tau_S)d\tau_S\right)d\tau_A\left(1 - \int_0^t f_B^o(\tau_B)d\tau_B\right) \\
&+ \int_0^t g_{B_L}^o(\tau_B) \\
&\left(\int_0^{\tau_B} f_{S_G}^s(\tau_S) + \left(1 - \int_0^{\tau_A} f_S^s(\tau_S)d\tau_S\right)\int_{\tau_A}^t f_{S_G}^o(\tau_S - \tau_B + \gamma\tau_S)d\tau_S\right)d\tau_B\cdot\left(1 - \int_0^t f_A^o(\tau_A)d\tau_A\right) \\
&+ \int_0^t f_{S_G}^s(\tau_S)d\tau_S\left(1 - \int_0^t f_A^o(\tau_A)d\tau_A\right)\left(1 - \int_0^t f_B^o(\tau_B)d\tau_B\right).
\end{aligned}
\tag{5.10}
$$

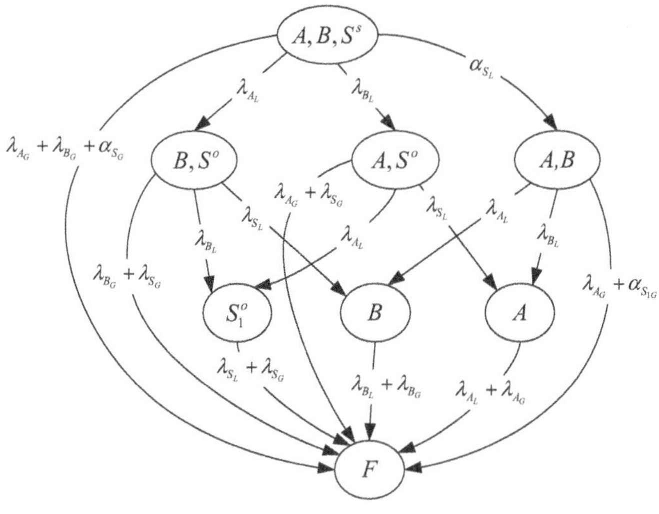

Fig. 5.13 Markov Model for the hard disk example

Results of the proposed method are compared to those obtained using the Markov model, which is shown in Fig. 5.13 for the case of exponential time-to-failure distributions.

Analysis of this example uses the following parameter values for the two primary and the one spare hard drives: $\lambda_{AL} = 0.09$/day, $\lambda_{AG} = 0.01$/day, $\lambda_{BL} = 0.18$/day, $\lambda_{BG} = 0.02$/day, $\alpha_{SG} = 5 \times 10^{-3}$/day, $\lambda_{SG} = 0.03$/day, $\alpha_{SL} = 0.045$/day, $\lambda_{SL} = 0.27$/day. As shown in Table 5.2, the results using both methods match exactly for different mission times.

- We also analyze the example using Lognormal and Weibull distributions to demonstrate the applicability of the SMDD method to different types of distributions. For the lognormal distribution case, the *pdf* function f is given by $\frac{1}{x\sigma\sqrt{2\pi}} e^{-(\ln x - \mu)^2/2\sigma^2}$, where μ and σ denote the mean and variance, respectively. In the case of Weibull distribution, the *pdf* function f is given by $(k/\lambda)(x/\lambda)^{k-1}e^{-(x/\lambda)^k}$, where k and λ denote the shape and scale parameters, respectively.
- For the lognormal distribution case, the parameters used for analysis are: $\mu\lambda_{AL} = 4$, $\mu\lambda_{AG} = 5$; $\mu\lambda_{BL} = 7.5$, $\mu\lambda_{BG} = 6$; $\mu\alpha_{SG} = 10$, $\mu\lambda_{SG} = 11$, $\mu\alpha_{SL} = 6$, $\mu\lambda_{SL} =$

Table 5.2 Results of both Markov and SMDD methods

Time	Markov	Proposed Method
5	0.26885	0.26885
10	0.5755	0.5755
15	0.76815	0.76815

Table 5.3 System unreliability results

Distribution time (Days)	Lognormal	Weibull
5	0.68223	0.42994
15	0.71553	0.53229
25	0.78322	0.64678

8.5, $\sigma\lambda_{AL} = 4$, $\sigma\lambda_{AG} = 8$, $\sigma\lambda_B = 7.5$, $\sigma\alpha_{SG} = 8$, $\sigma\lambda_{SG} = 8.8$, $\sigma\alpha_{SL} = 3$, $\sigma\lambda_{SL} = 6$ and $\varphi_s = 1$.

- For the Weibull distribution case, the parameter values used for analysis are: $k_{AL} = k_{AG} = 2$, $k_{BL} = k_{BG} = 0.9$, $k_{S\alpha G} = k_{S\alpha L} = 1.1$, $k_{S\alpha G} = k_{S\lambda G} = 0.9$, $\lambda_{AL} = 0.09$/day, $\lambda_{AG} = 0.01$/day, $\lambda_{BL} = 0.18$/day, $\lambda_{AG} = 0.02$/day, $\alpha_{SG} = 0.005$/day, $\lambda_{SG} = 0.02$/day, $\alpha_{SL} = 0.045$/day, $\lambda_{SL} = 0.02$/day, $\varphi_s = 1$.
- Table 5.3 summarizes the final results for both cases.

5.2.2 *Case Study Two*

Figure 5.14 gives the DFT model where components *A* and *B* share the same warm spare *S*. The two primary components are working in series. The system fails when one of WSP gates fire. The spare *S* fails with a reduced failure rate α_s any time before the failure of any of the primary components. Otherwise if any of the primary components *A* or *B* fails first, the spare component *S* replaces it and can then fail with a full failure rate λ_s.

Similar to the previous examples, the DFT is converted to an OR gate connecting different events contributing to the system failure: any component failing globally or two components failing locally. This is illustrated in Fig. 5.15. Applying the tradition and extended SMDD rules, the SMDD is generated in the bottom-up manner from the converted DFT. Figure 5.16 shows the final system SMDD model.

There are 9 paths from the root to the sink node "1" after the merging:

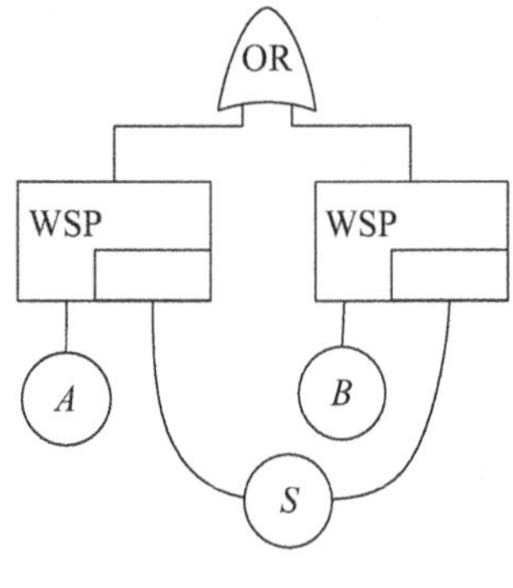

Fig. 5.14 The DFT of the system

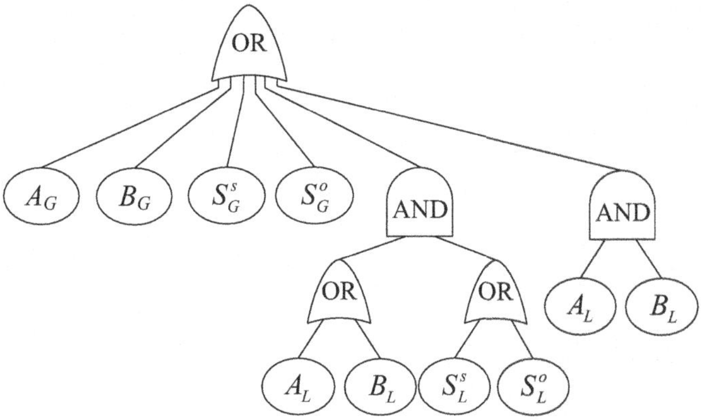

Fig. 5.15 The Converted DFT

Fig. 5.16 Final SMDD in case study two

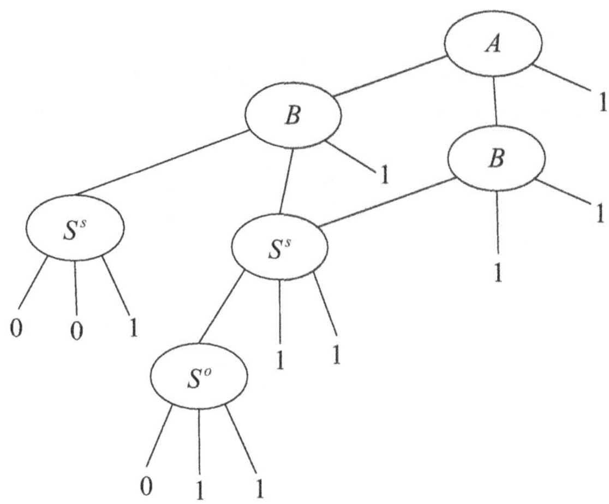

1: A_G;	2: $A_L \sim B_G$;	3: $A \sim B_G$;
4: $A_L \sim B_L$;	5: $A_L \sim B \sim S^s$;	6: $A_L \sim B \sim S^o$;
7: $A \sim B_L \sim S^s$;	8: $A \sim B_L \sim S^o$;	9: $A \sim B \sim S_G^s$

The second and third paths are further merged into one path: $A_G \sim B_G$. The system unreliability is the sum of probabilities of all the 9 paths as:

$$
\begin{aligned}
F_S(t) &= \int_0^t f_{A_G}^o(\tau_A)d\tau_A + \left(1 - \int_0^t f_{A_G}^o(\tau_A)d\tau_A\right)\int_0^t f_{B_G}^o(\tau_B)d\tau_B \\
&+ \int_0^t g_{A_L}^o(\tau_A)d\tau_A \int_0^t f_B^o(\tau_B)d\tau_B \\
&+ \int_0^t g_{A_L}^o(\tau_A)d\tau_A \left(1 - \int_0^{\tau_A} g_{B_L}^o(\tau_B)d\tau_B\right) \\
&\left[\int_0^{\tau_A} f_S^s(\tau_S)d\tau_S + \left(1 - \int_0^{\tau_A} f_S^s(\tau_S)d\tau_S\right)\int_{\tau_A}^t f_S^o(\tau_S - \tau_A + \gamma\tau_A)d\tau_S\right]d\tau_B d\tau_A \\
&+ \int_0^t g_{A_L}^o(\tau_A)d\tau_A \int_0^t f_B^o(\tau_B)d\tau_B \\
&+ \left(1 - \int_0^t g_{A_L}^o(\tau_A)d\tau_A\right)\left(1 - \int_0^t f_B^o(\tau_B)d\tau_B\right)\int_0^t f_{S_G}^s(\tau_S)d\tau_S
\end{aligned}
\tag{5.11}
$$

Results of the proposed method are compared to those obtained using the Markov model, which is shown in Fig. 5.17 for the case of exponential time-to-failure distributions. Analysis of this example uses the following parameter values for the two primary and the one spare components:

$\lambda_{AL} = 0.09$/day, $\lambda_{AG} = 0.01$/day, $\lambda_{AL} = 0.18$/day, $\lambda_{AG} = 0.02$/day, $\alpha_{SG} = 5 \times 10^{-3}$/day, $\lambda_{SG} = 0.03$/day, $\alpha_{SL} = 0.045$/day, $\lambda_{SL} = 0.27$/day. As shown in Table 5.4, the results using both methods match exactly for different mission times.

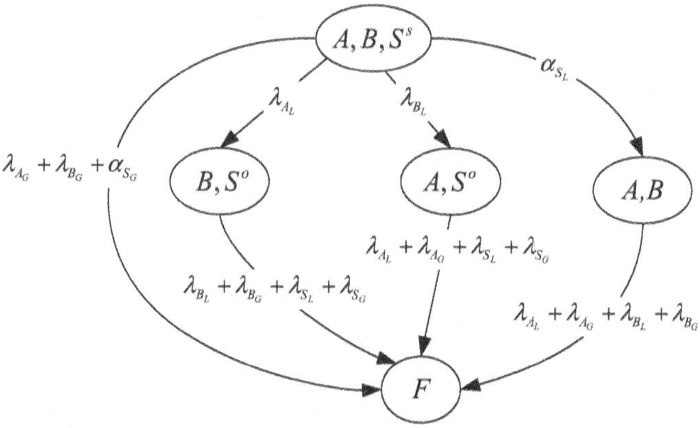

Fig. 5.17 Markov Model for case study two

Table 5.4 Results of both Markov and SMDD methods

Time (Days)	Markov	Proposed Method
5	0.58841	0.58841
10	0.89532	0.89532
15	0.97669	0.97669

5.3 Imperfect Switching

The previous sections assume perfect switching. However, in real applications the switching mechanism in standby systems is not perfectly reliably and can fail, leading to the system failure even if the standby units are operational. Based on the total probability law, the system failure probability is sum of the probability of primary components and standby components failures that lead to the system failure given perfect switching and the probability of switch failure without being able to perform its switching function after detecting a fault in the primary unit. The system unreliability is then given by:

$$\begin{aligned} UR = &\Pr\{\text{System failure}|\text{perfect switch}\} \cdot \Pr\{\text{switch operational }\} \\ &+ \Pr\{\text{System failure imperfect switch}\} \cdot \Pr\{\text{Switch failure}\} \end{aligned} \tag{5.12}$$

The first term is the system unreliability assuming perfect switch; the second term is the combination of switch and all primary unit's failure requiring switchover.

The switch failure probability depends on the switching model. The switch failure probability can be either constant at switching time. Or, it can be age dependent meaning the switch can fail anytime like other components in the system, and its failure probability depends on the time. Despite the switch model, the imperfect switching is incorporated into the SMDD by adding n switch nodes (n is the number of standby components), one after each local failure, specifically between the two nodes S_i^s and S_i^o as shown in Fig. 5.18. The switch can either fail locally which leads to the system failure in the case of the online component failure or functions correctly to switch the standby component into the operation.

For illustration, consider a WSP system with a primary unit P and one standby unit S. Figure 5.19 illustrates the SMDD assuming perfect switching.

In this example there is only one standby unit, hence one switch node is inserted after the node S_i^s. Figure 5.20 shows the SMDD with the incorporation of the imperfect switching.

The system unreliability is the sum of probability of all paths that lead to the sink node '1'.

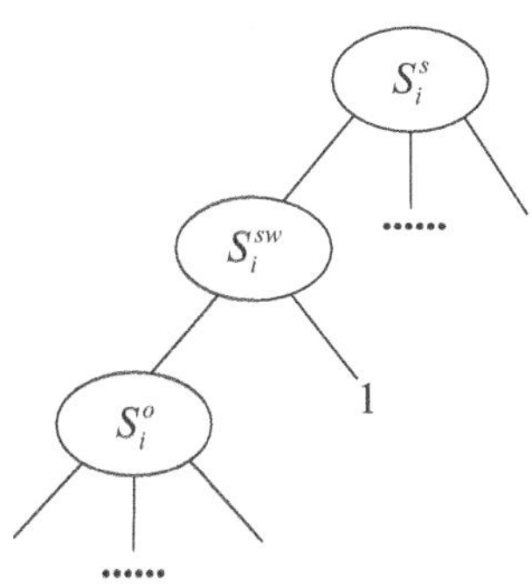

Fig. 5.18 Inserting a switch node to SMDD

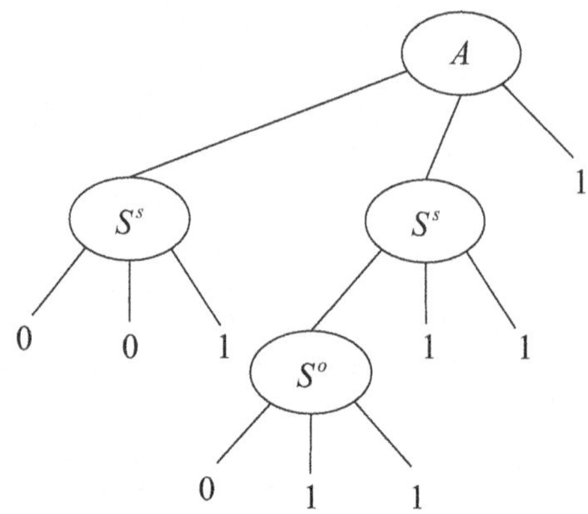

Fig. 5.19 SMDD in the case of perfect switching

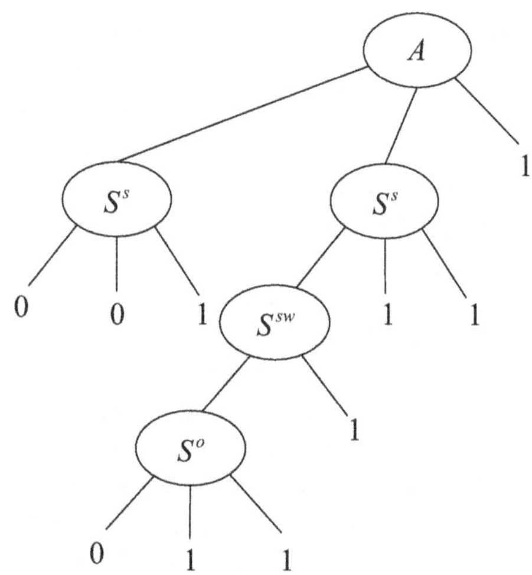

Fig. 5.20 SMDD in the case of imperfect switch

1:A_G;	2:$A_L \sim S^s_L$;	3:$A_L \sim S^s_G$;	4:$A_L \sim S^{sw}$;
5:$A_L \sim S^{sw} \sim S^o_L$;	6:$A_L \sim S^{sw} \sim S^o_G(\lambda)$;	7:$A \sim S^s_G$	

Similar to the previous examples, paths 2 and 3 can be merged to $A_L \gg S^s$, and paths 5 and 6 are merged into $A_L \sim S^{sw} \sim S^o$. The system unreliability is then given by

$$F_S(t) = \int_0^t g^o_{A_L}(\tau_1)\left(\begin{array}{l} \int_0^{\tau_1} f^s_S(\tau_2)d\tau_2 + \left(1 - \int_0^{\tau_1} f^s_S(\tau_3)d_3\right)\int_0^{\tau_1} f^{sw}_S(\tau_4)d\tau_4 \\ +\left(1 - \int_0^{\tau_1} f^{sw}_S(\tau_4)d\tau_4\right)\cdot\left(1 - \int_0^{\tau_1} f^s_S(\tau_3)d\tau_3\right)\int_{\tau_1}^t f^o_S(\tau_2 - \tau_1 + \gamma\tau_2)d\tau_2 \end{array}\right)d\tau_1$$
$$+\int_0^t f^o_{A_G}(\tau_1)d\tau_1 + \int_0^t f^s_{S_G}(\tau_2)d\tau_2\left(1 - \int_0^t f^o_A(\tau_1)d\tau_1\right) \quad (5.13)$$

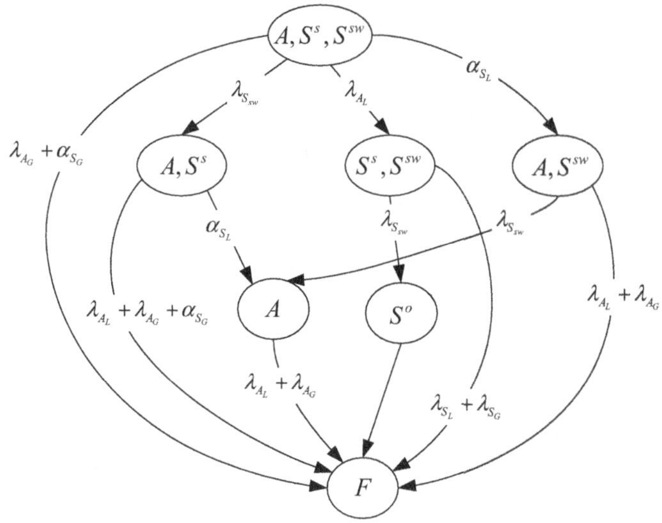

Fig. 5.21 Markov Model for WSP with switch failure

Table 5.5 Results of both Markov and SMDD methods

Time (Days)	Markov	Proposed Method
5	0.26451	0.26451
10	0.3475	0.3475
15	0.5793	0.5793

Again, the results of the proposed method are compared to those obtained by the Markov model in Fig. 5.21. Table 5.5 shows results using both the proposed method and the Markov method, which exactly match. The following parameter values for the primary and standby spares are used: $\lambda_{AL} = 0.18$/day, $\lambda_{AG} = 0.02$/day, $\alpha_{SG} = 0.005$/day, $\alpha_{SL} = 0.045$/day, $\lambda_{SG} = 0.01$/day, $\lambda_{SL} = 0.09$/day. For switch, the failure rate is $\lambda_{Sw} = 0.01$/day.

References

1. Akers J, Bergman R, Amari S V, et al. Analysis of multi-state systems using multi-valued decision diagrams [A]. Annual Reliability and Maintainability Symposium (RAMS2008), Las Vegas, NV, 2008 [C].Piscataway, NJ: IEEE, 347–353.
2. Amari S V, Xing L, Shrestha A, et al. Performability analysis of multistate computing systems using multivalued decision diagrams [J]. IEEE Transactions on Computers, 2010, 9(10): 1419–1433.
3. Xing L, Dai Y. A new decision diagram based method for efficient analysis on multi-state systems [J]. IEEE Transactions on Dependable and Secure Computing, 2009, 6(3): 161–174.

4. Xing L, Levitin G. Combinatorial algorithm for reliability analysis of multistate systems with propagated failures and failure isolation effect [J]. IEEE Transactions on Systems Man and Cybernetics-Part A: Systems and Humans, 2011, 41(6): 1156–1165.
5. Solomon P. Effect of misspecification of regression models in the analysis of survival data [J]. Baometrzka, 1984, 71(2): 291–298.
6. Misra K B (Editor). Handbook of Performability Engineering, Springer-Verlag, London, ISBN: 978–1–84800–130–5, 2008.
7. Nelson W. Accelerated life testing step-stress models and data analysis [J]. IEEE Transaction on Reliability, 1980, 29: 103–108.
8. DeGroot M, Goel P. Bayesian estimation and optimal designs in partially accelerated life testing [J]. Naval Research Logistics Quarterly, 1979, 26: 223–235.
9. Nelson W. Accelerated Testing: Statistical Models, Test Plans, and Data Analyses. Wiley & Sons, 1990.
10. Bhattacharyya G, Soejoeti Z. A tampered failure rate model for step-stress accelerated life test [J].Communications in Statistics-Theory and Methods, 1989, 18(5): 1627–1643.
11. Amari S V, Bergman R. Reliability analysis of k-out-of-n load-sharing systems [A]. Annual Reliability & Maintainability Symposium (RAMS2008), LA, USA, 2008 [C]. Piscataway, NJ: IEEE, 440–445.

Chapter 6
Optimal Working Sequence in a 1-Out-Of-N Warm Standby System

In a warm standby system, the components are working following a predetermined order. At the beginning, the primary components are at the normal working state to sustain the system operation. When some of the primary components fail, the warm standby components would be switched to the online state. Then, a nature question is which components should be used as the primary components and which component should be used as standby components. Apparently, if all the components are identical, then the working order would not be relevant. In contrast, the system reliability may be different if the components have different reliability characteristics. In this case, the working order of components is relevant to the system reliability.

For the hot standby system, all the components work in parallel, and the working order seems irrelevant. For the cold standby system, the system lifetime equals the sum of the lifetime of the primary component and the standby component, and the working order is also irrelevant. In these cases, the working order has no impact on the system reliability. However, the system lifetime has a more complicated dependency on the lifetimes of its components, and it is not clear at the moment whether the system reliability is dependent on different working orders of the components.

6.1 System Reliability for a 1-Out-Of-n Warm Standby System

Denote the CDF of a component under the normal working state by $F^o(t)$ and that under the warm standby state by $F^s(t)$. In general, because the working load in the warm standby state is milder than the normal working state, $F^s(t)$ is smaller than $F^o(t)$. We use the AFTM to characterize the lifetimes of a component under the warm standby state and the normal working state [1, 2]. In AFTM, the milder stress has the effect of expanding time through a factor $0 < \gamma < 1$, which can be expressed as

R. Peng et al., *Reliability Modelling and Optimization of Warm Standby Systems*,
https://doi.org/10.1007/978-981-16-1792-8_6

$$F^s(t) = F^o(\gamma t), \tag{6.1}$$

A more general model is

$$F^s(t) = F^o(\gamma(t)), \tag{6.2}$$

where $\gamma(t) \leq t$ is a non-decreasing function for $t > 0$, and $\gamma(0) = 0$.

A warm standby component would first stay in the warm standby state and then may be switched to the normal working state at some time τ. Based on the AFTM and CEM as discussed in Sect. 2.1, the lifetime distribution for the warm standby component is

$$F(t) = \begin{cases} F^o(\gamma(t)), 0 < t \leq \tau \\ F^o(t - \tau + \gamma(\tau)), t > \tau \end{cases}, \tag{6.3}$$

Thus, the system lifetime can be characterized by $F^o(t)$ given $\gamma(t)$. In the following, we will use $F(t)$ without superscript "o" to denote the CDF of a component at the normal working state, and assume that $\gamma(t)$ is the same for all the components even if different components may have different lifetime distributions.

For a 1-out-of-2 warm standby system with primary component A_1 and standby component A_2, the system reliability function can be written as

$$\begin{aligned} R_S^{1,2}(t) &= \Pr\{A_1 \text{ is operational at } t\} \\ &\quad + \Pr\{A_1 \text{ fails before } t \& A_2 \text{ is operational at } t\} \\ &= R_1(t) + \int_0^t R_2(t - \tau + \gamma(\tau)) f_1(\tau) d\tau, \end{aligned} \tag{6.4}$$

where the superscript of $R_S^{1,2}(t)$ indicates that component A_1 is the primary online component and A_2 is the warm standby one. Similarly, the CDF of the system lifetime is

$$F_S^{1,2}(t) = 1 - R_S^{1,2}(t) = \int_0^t F_2(t - \tau + \gamma(\tau)) f_1(\tau) d\tau. \tag{6.5}$$

From (6.5), the PDF of the system lifetime can be obtained as

$$\begin{aligned} f_S^{1,2}(t) &= F_2(\gamma(t)) f_1(t) + \int_0^t f_2(t - \tau + \gamma(\tau)) f_1(\tau) d\tau \\ &= f_1(t) - R_2(\gamma(t)) f_1(t) + \int_0^t f_2(t)(t - \tau + \gamma(\tau)) f_1(\tau) d\tau \end{aligned}. \tag{6.6}$$

For a 1-out-of-3 warm standby system with primary component A_1, and standby components A_2, A_3, the system reliability can be obtained as

$$
\begin{aligned}
R_S^{1,2,3}(t) &= \Pr\{A_1 \text{ is operational at } t\} \\
&\quad + \Pr\{A_1 \text{ fails before } t \,\&\, A_2 \text{ is operational at } t\} \\
&\quad + \Pr\{\text{Both } A_1, A_2 \text{ fail before } t \,\&\, A_3 \text{ is operational at } t\} \\
&= \Pr\{\text{At least one of } A_1, A_2 \text{ is operational at } t\} \\
&\quad + \Pr\{\text{Both } A_1, A_2 \text{ fail before } t \,\&\, A_3 \text{ is operational at } t\} \\
&= R_S^{1,2}(t) + \int_0^t R_3(t-\tau+\gamma(\tau)) f_S^{1,2}(\tau)d\tau \\
&= R_1(t) + \int_0^t R_2(t-\tau+\gamma(\tau)) f_1(\tau)d\tau \\
&\quad + \int_0^t R_3(t-\tau+\gamma(\tau)) f_1(\tau)d\tau \\
&\quad + \int_0^t R_3(t-\tau+\gamma(\tau)) \left(\int_0^\tau f_2(\tau-u+\gamma(u)) f_1(u)du - R_2(\gamma(\tau)) f_1(\tau) \right) d\tau,
\end{aligned} \tag{6.7}
$$

Generally, for a 1-out-of-n warm standby system, the system reliability can be obtained with the recursive formula

$$
R_S^{1,2,\ldots,n}(t) = R_S^{1,2,\ldots,(n-1)}(t) + \int_0^t R_n(t-\tau+\gamma(\tau)) f_S^{1,2,\ldots,(n-1)}(\tau)d\tau, \tag{6.8}
$$

6.2 Optimal Component Working Order for a Two-Component System

We first consider the optimal component working order in a two-component warm standby system. We consider the system reliability and the expected lifetime as two objectives in the design of the component working order and investigate the optimal working order that maximizes these objectives.

6.2.1 Optimal Working Component Order Maximizing the Expected System Lifetime

First consider the expected system lifetime as the objective. Cha et al. [3] proved that for a 1-out-of-2 warm standby system with exponential components and linear AFTM, *i.e.*$\gamma(t) = \gamma t$, if the failure rate of A_1 is larger than that of A_2, $\lambda_1 > \lambda_2$, then the expected lifetime of the system with A_1 as the primary component $E\left[T_S^{1,2}\right]$ is larger than that with A_2 as the primary component $E\left[T_S^{2,1}\right]$. Here, T_S^{*} represents the system lifetime with structure "*". This implies that the weaker component with higher failure rate should be used first in terms of the expected system lifetime. In the following, we show that this result can be extended to a more general case with nonlinear AFTM and general distributions for components.

Theorem 6.1. Suppose the function $\gamma(t)$ is strictly increasing, $\gamma(0) = 0$, $\gamma(t) < t, \forall t > 0$, $\lim_{t\to+\infty} \gamma(t) = +\infty$. If the failure rate functions of the two components satisfy $\lambda_1(t) > \lambda_2(t), \forall t > 0$, then we have

$$E\left[T_S^{1,2}\right] > E\left[T_S^{2,1}\right].$$

Proof. We first consider $E\left[T_S^{1,2}\right]$. Here,

$$\begin{aligned}
E\left[T_S^{1,2}\right] &= \int_0^{+\infty} R_S^{1,2}(t)dt = \int_0^{+\infty}\left(R_1(t) + \int_0^t R_2(t-u+\gamma(u))f_1(u)du\right)dt \\
&= \int_0^{+\infty} R_1(t)dt + \int_0^{+\infty}\int_0^t R_2(t-u+\gamma(u))f_1(u)dudt \\
&\underset{w=t-u+\gamma(u)}{\overset{v=u}{=}} E[T_1] + \int_0^{+\infty}\int_0^{\gamma^{-1}(w)} R_2(w)f_1(v)dvdw \\
&= E[T_1] + \int_0^{+\infty} R_2(w)F_1\left(\gamma^{-1}(w)\right)dw \\
&= E[T_1] + E[T_2] - \int_0^{+\infty} R_2(w)R_1\left(\gamma^{-1}(w)\right)dw,
\end{aligned} \tag{6.9}$$

where $\gamma^{-1}(\cdot)$ represents the inverse function of $\gamma(\cdot)$. Note that, for the change of variables here

$$\begin{pmatrix} v \\ w \end{pmatrix} = \begin{pmatrix} u \\ t-u+\gamma(u) \end{pmatrix},$$

the Jacobian matrix and the determinant are

$$\begin{pmatrix} 1 & 0 \\ \gamma'(u)-1 & 1 \end{pmatrix} \text{and} 1.$$

Similarly, we can derive that

$$E\left[T_S^{2,1}\right] = E[T_1] + E[T_2] - \int_0^{+\infty} R_1(w)R_2\left(\gamma^{-1}(w)\right)dw. \tag{6.10}$$

Comparing (6.9) and (6.10), we can see that $E\left[T_S^{1,2}\right] > E\left[T_S^{2,1}\right]$ if $R_2(w)R_1\left(\gamma^{-1}(w)\right) < R_1(w)R_2\left(\gamma^{-1}(w)\right)$ holds for all $w > 0$.

According to the basic relationship $R(t) = \exp\left\{-\int_0^t \lambda(\tau)d\tau\right\}$, we have

$$\begin{aligned}
& R_2(w)R_1\left(\gamma^{-1}(w)\right) < R_1(w)R_2\left(\gamma^{-1}(w)\right) \\
\Leftrightarrow\ & \frac{\exp\left\{-\int_0^{\gamma^{-1}(w)} \lambda_1(\tau)d\tau\right\}}{\exp\left\{-\int_0^{w} \lambda_1(\tau)d\tau\right\}} < \frac{\exp\left\{-\int_0^{\gamma^{-1}(w)} \lambda_2(\tau)d\tau\right\}}{\exp\left\{-\int_0^{w} \lambda_2(\tau)d\tau\right\}} \\
\Leftrightarrow\ & \exp\left\{-\int_w^{\gamma^{-1}(w)} \lambda_1(\tau)d\tau\right\} < \exp\left\{-\int_w^{\gamma^{-1}(w)} \lambda_2(\tau)d\tau\right\} \\
\Leftrightarrow\ & \int_w^{\gamma^{-1}(w)} \lambda_1(\tau)d\tau > \int_w^{\gamma^{-1}(w)} \lambda_2(\tau)d\tau.
\end{aligned} \tag{6.11}$$

Because $\gamma(t)$ is a strictly monotonically increasing function, $\gamma(t) < t$ indicates $w < \gamma^{-1}(w)$. We also know that $\lambda_1(t) > \lambda_2(t)$, which indicates that $\int_w^{\gamma^{-1}(w)} \lambda_1(\tau)d\tau > \int_w^{\gamma^{-1}(w)} \lambda_2(\tau)d\tau$. This completes the proof.

Example 6.1. Suppose the failure rate functions of components A_1 and A_2 are $\lambda_1(t) = 0.01 + 2 \times 10^{-4}t$ and $\lambda_2(t) = 2 \times 10^{-4}t$, respectively, and $\gamma(t) = 0.5t$. Then we have $E\left[T_S^{1,2}\right] = 117.671$ and $E\left[T_S^{2,1}\right] = 111.862$. It is obvious that using the weaker component first significantly increases the expected system lifetime, and thus improves the system reliability.

6.2.2 *Optimal Working Component Order Maximizing the System Reliability*

Still consider the two-component warm standby system. For a special case of Theorem 6.1 where the function $\gamma(t)$ satisfies the linear AFTM, i.e., $\gamma(t) = \gamma t, 0 < \gamma < 1$, a stronger result could be obtained in Theorem 6.2. To proof Theorem 6.2, we shall first introduce the following Lemma 6.1 is introduced to prove Theorem 3, and then Theorem 3 follows.

Lemma 6.1. Function

$$h(x) = \frac{x\left(1 - x^{n-2}\right)}{1 - x^n} = \frac{x - x^{n-1}}{1 - x^n} (n \geq 3)$$

is strictly increasing on (0, 1).

Proof. The derivative of function $h(x)$ could be obtained directly:

$$\frac{dh(x)}{dx} = \frac{1 - x^{2n-2} - (n-1)x^{n-2}\left(1 - x^2\right)}{\left(1 - x^n\right)^2}$$

where

$$
\begin{aligned}
&1-x^{2n-2}-(n-1)x^{n-2}(1-x^2)\\
&=(1-x^2)\sum_{k=0}^{n-2}(x^2)^k-(1-x^2)(n-1)x^{n-2}\\
&=(1-x^2)(n-1)\left(\frac{1}{n-1}\sum_{k=0}^{n-2}(x^2)^k-x^{n-2}\right).
\end{aligned}
$$

With the well known inequality of arithmetic and geometric means, we have

$$\frac{1}{n-1}\sum\nolimits_{k=0}^{n-2}(x^2)^k\geq\left(\prod\nolimits_{k=0}^{n-2}(x^2)^k\right)^{n-1}=\left(x^{2\sum_{k=0}^{n-2}k}\right)^{\frac{1}{n-1}}=x^{n-2},$$

and the inequality strictly holds for $x\in(0,1)$. So $dh(x)/dx>0$ holds for $x\in(0,1)$, which implies $h(x)$ is strictly increasing for $x\in(0,1)$.

Theorem 6.2. Suppose ATFM is linear with the function $\gamma(t)=\gamma t,\gamma\in(0,1)$. For a 1-out-of-2 warm standby system with exponential components, if the failure rates of two components satisfy $\lambda_1>\lambda_2$, then we have.

$$R_S^{1,2}(t)>R_S^{2,1}(t),\forall t>0.$$

Proof. According to (6. 4), the system reliability for this 1-out-of-2 warm standby system is

$$
\begin{aligned}
R_S^{1,2}(t)&=R_1(t)+\int_0^t R_2(t-\tau+\gamma\tau)f_1(\tau)d\tau\\
&=e^{-\lambda_1 t}+\int_0^t e^{-\lambda_2(t-\tau+\gamma\tau)}\cdot\lambda_1 e^{-\lambda_1\tau}d\tau\\
&\overset{\delta=1-\gamma}{\Rightarrow}e^{-\lambda_1 t}+\frac{\lambda_1 e^{-(\lambda_1+\lambda_2)t}\left(e^{\lambda_1 t}-e^{\delta\lambda_2 t}\right)}{\lambda_1-\delta\lambda_2}\\
&=e^{-\lambda_1 t}+\frac{\lambda_1\left(e^{-\lambda_2 t}-e^{-\lambda_1 t-(1-\delta)\lambda_2 t}\right)}{\lambda_1-\delta\lambda_2},
\end{aligned}
$$

where $\delta=1-\gamma\in(0,1)$.
Similarly we have

$$R_S^{2,1}(t)=\begin{cases}e^{-\lambda_2 t}+\frac{\lambda_2\left(e^{-\lambda_1 t}-e^{-\lambda_2 t-(1-\delta)\lambda_1 t}\right)}{\lambda_2-\delta\lambda_1},\lambda_2\neq\delta\lambda_1\\ e^{-\lambda_2 t}+\lambda_2 t\cdot e^{-\lambda_1 t},\lambda_2=\delta\lambda_1\end{cases}.$$

In the following, we prove that $R_S^{1,2}(t)>R_S^{2,1}(t)$ for $\lambda_2\neq\delta\lambda_1$ and $\lambda_2=\delta\lambda_1$.

(1) $\lambda_2 \neq \delta\lambda_1$.

In this case, the following relationship holds:

$$
\begin{aligned}
& R_S^{1,2}(t) > R_S^{2,1}(t) \\
& \Leftrightarrow e^{-\lambda_1 t} + \frac{\lambda_1\left(e^{-\lambda_2 t} - e^{-\lambda_1 t-(1-\delta)\lambda_2 t}\right)}{\lambda_1 - \delta\lambda_2} > e^{-\lambda_2 t} \\
& \quad + \frac{\lambda_2\left(e^{-\lambda_1 t} - e^{-\lambda_2 t-(1-\delta)\lambda_1 t}\right)}{\lambda_2 - \delta\lambda_1} \Leftrightarrow \frac{\delta\lambda_1^2\left(e^{\lambda_2 t} - e^{\delta\lambda_2 t}\right) + \lambda_1\lambda_2\left(e^{\delta\lambda_2 t} - \delta^2 e^{\lambda_2 t}\right)}{(\lambda_2 - \delta\lambda_1)(\lambda_1 - \delta\lambda_2)} \\
& \quad < \frac{\delta\lambda_2^2\left(e^{\lambda_1 t} - e^{\delta\lambda_1 t}\right) + \lambda_1\lambda_2\left(e^{\delta\lambda_1 t} - \delta^2 e^{\lambda_1 t}\right)}{(\lambda_2 - \delta\lambda_1)(\lambda_1 - \delta\lambda_2)}.
\end{aligned} \tag{6.12}
$$

With the power series expansion $e^x = \sum_{n=0}^{+\infty}\frac{1}{n!}x^n$, we have

$$
\begin{aligned}
& \delta\lambda_1^2\left(e^{\lambda_2 t} - e^{\delta\lambda_2 t}\right) + \lambda_1\lambda_2\left(e^{\delta\lambda_2 t} - \delta^2 e^{\lambda_2 t}\right) \\
& = \delta\lambda_1^2\left(\sum_{n=0}^{\infty}\frac{1}{n!}\left((\lambda_2 t)^n - (\delta\lambda_2 t)^n\right)\right) + \lambda_1\lambda_2\left(\sum_{n=0}^{\infty}\frac{1}{n!}\left((\delta\lambda_2 t)^n - \delta^2(\lambda_2 t)^n\right)\right) \\
& = \sum_{n=0}^{\infty}\frac{1}{n!}t^n\left(\delta\lambda_1^2\lambda_2^n\left(1-\delta^n\right) + \lambda_1\lambda_2^{n+1}\left(\delta^n - \delta^2\right)\right).
\end{aligned}
$$

Therefore, we have

$$
\begin{aligned}
& R_S^{1,2}(t) > R_S^{2,1}(t) \\
& \Leftrightarrow \frac{\delta\lambda_1^2\left(e^{\lambda_2 t} - e^{\delta\lambda_2 t}\right) + \lambda_1\lambda_2\left(e^{\delta\lambda_2 t} - \delta^2 e^{\lambda_2 t}\right) - \delta\lambda_2^2\left(e^{\lambda_2 t} - e^{\delta\lambda_1 t}\right) - \lambda_1\lambda_2\left(e^{\delta\lambda_1 t} - \delta^2 e^{\lambda_1 t}\right)}{(\lambda_2 - \delta\lambda_1)(\lambda_1 - \delta\lambda_2)} < 0 \\
& \Leftrightarrow \frac{\sum_{n=0}^{\infty}\frac{1}{n!}t^n\left(\delta\lambda_1^2\lambda_2^n(1-\delta^n) + \lambda_1\lambda_2^{n+1}\left(\delta^n - \delta^2\right)\right) - \sum_{n=0}^{\infty}\frac{1}{n!}t^n\left(\delta\lambda_2^2\lambda_1^n(1-\delta^n) + \lambda_2\lambda_1^{n+1}\left(\delta^n - \delta^2\right)\right)}{(\lambda_2 - \delta\lambda_1)(\lambda_1 - \delta\lambda_2)} < 0 \\
& \Leftrightarrow \frac{\sum_{n=3}^{\infty}\frac{1}{n!}t^n\left(\delta\lambda_1^2\lambda_2^n(1-\delta^n) + \lambda_1\lambda_2^{n+1}\left(\delta^n - \delta^2\right) - \delta\lambda_2^2\lambda_1^n(1-\delta^n) - \lambda_2\lambda_1^{n+1}\left(\delta^n - \delta^2\right)\right)}{(\lambda_2 - \delta\lambda_1)(\lambda_1 - \delta\lambda_2)} < 0 \\
& \overset{k=\frac{\lambda_2}{\lambda_1}}{\Leftrightarrow} \frac{\sum_{n=3}^{\infty}\frac{1}{n!}t^n\lambda_1^n\left(\delta k^n(1-\delta^n) + k^{n+1}\left(\delta^n - \delta^2\right) - \delta k^2(1-\delta^n) - k\left(\delta^n - \delta^2\right)\right)}{(k-\delta)(1-\delta k)} < 0 \\
& \Leftrightarrow \frac{\sum_{n=3}^{\infty}\frac{1}{n!}t^n\lambda_1^n\delta k(1-k^n)(1-\delta^n)\left(\frac{\delta\left(1-\delta^{n-2}\right)}{1-\delta^n} - \frac{k\left(1-k^{n-2}\right)}{1-k^n}\right)}{(k-\delta)(1-\delta k)} > 0,
\end{aligned}
$$

where $0 < k = \lambda_2/\lambda_1 < 1$.

According to Lemma 6.1,

$$
h(x) = \frac{x\left(1-x^{n-2}\right)}{1-x^n} (n \geq 3)
$$

is strictly increasing for $x \in (0, 1)$. Therefore,

$$\frac{\delta(1-\delta^{n-2})}{1-\delta^n} - \frac{k(1-k^{n-2})}{1-k^n} \sim \begin{cases} > 0, \delta - k > 0 \\ < 0, \delta - k < 0 \end{cases}.$$

Subsequently,

$$\frac{\sum_{n=3}^{\infty} \frac{1}{n!} t^n \lambda_1^n \delta k (1-k^n)(1-\delta^n)\left(\frac{\delta(1-\delta^{n-2})}{1-\delta^n} - \frac{k(1-k^{n-2})}{1-k^n}\right)}{(k-\delta)(1-\delta k)} > 0, \forall t > 0.$$

That is, $R_S^{1,2}(t) > R_S^{2,1}(t), \forall t > 0$ holds for $\lambda_2 \neq \delta\lambda_1$.

(2) $\lambda_2 = \delta\lambda_1$.

For the case $\lambda_2 = \delta\lambda_1$, the following derivations hold:

$$\begin{aligned}
& R_S^{1,2}(t) > R_S^{2,1}(t) \\
& \Leftrightarrow e^{-\lambda_1 t} + \frac{\lambda_1\left(e^{-\lambda_2 t} - e^{-\lambda_1 t - (1-\delta)\lambda_2 t}\right)}{\lambda_1 - \delta\lambda_2} > e^{-\lambda_2 t} + \lambda_2 t e^{-\lambda_1 t} \\
& \Leftrightarrow e^{-\lambda_1 t} + \frac{\lambda_1\left(e^{-\delta\lambda_1 t} - e^{-\lambda_1 t - (1-\delta)\delta\lambda_1 t}\right)}{\lambda_1 - \delta^2\lambda_1} > e^{-\delta\lambda_1 t} + \delta\lambda_1 t e^{-\lambda_1 t} \\
& \Leftrightarrow e^{\delta\lambda_1 t} + \frac{\left(e^{\lambda_1 t} - e^{\delta^2\lambda_1 t}\right)}{1-\delta^2} > e^{\lambda_1 t} + \delta\lambda_1 t e^{\delta\lambda_1 t} \\
& \Leftrightarrow \left(1-\delta^2\right)e^{\delta\lambda_1 t} + \delta^2 e^{\lambda_1 t} - e^{\delta^2\lambda_1 t} > \left(1-\delta^2\right)\delta\lambda_1 t e^{\delta\lambda_1 t}.
\end{aligned} \tag{6.13}$$

With the power series expansion $e^x = \sum_{n=0}^{+\infty} \frac{1}{n!} x^n$, we have

$$\begin{aligned}
& R_S^{1,2}(t) > R_S^{2,1}(t) \\
& \Leftrightarrow \left(1-\delta^2\right)e^{\delta\lambda_1 t} + \delta^2\left(e^{\lambda_1 t} - e^{\delta^2\lambda_1 t}\right) > \left(1-\delta^2\right)\delta\lambda_1 t e^{\delta\lambda_1 t} \\
& \Leftrightarrow \sum_{n=0}^{\infty} \frac{1}{n!}(\lambda_1 t)^n\left(\left(1-\delta^2\right)\delta^n + \delta^2 - \delta^{2n}\right) - \sum_{n=1}^{\infty} \frac{1}{n!}(\lambda_1 t)^n n\left(1-\delta^2\right)\delta^n > 0 \\
& \Leftrightarrow \sum_{n=3}^{\infty} \frac{1}{n!}(\lambda_1 t)^n\left(\left(\delta^2 + \delta^n\right)\left(1-\delta^n\right) - n\left(1-\delta^2\right)\delta^n\right) > 0 \\
& \Leftrightarrow \sum_{n=3}^{\infty} \frac{1}{n!}(\lambda_1 t)^n \delta^2(1-\delta)\left(\left(1+\delta^{n-2}\right)\sum_{i=0}^{n-1} \delta^i - n(1+\delta)\delta^{n-2}\right) > 0 \\
& \Leftrightarrow \sum_{n=3}^{\infty} \frac{1}{n!}(\lambda_1 t)^n \delta^2(1-\delta)\left(\sum_{i=0}^{n-1} \delta^i\left(1-\delta^{n-1-i}\right) - \delta^{n-2}\sum_{i=0}^{n-1}\left(\delta^i - 1\right)\right) > 0
\end{aligned}$$

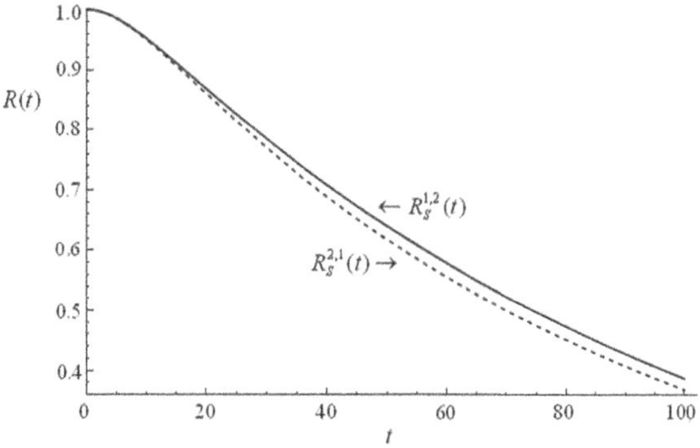

Figure. 6.1 Reliability functions of the 1-out-of-2 warm standby system with different component orders

$$\Leftrightarrow \sum_{n=3}^{\infty} \frac{1}{n!}(\lambda_1 t)^n \delta^2 (1-\delta)\left(\sum_{i=2}^{n-1} \delta^{n-1-i}\left(1-\delta^{i-1}\right)\left(1-\delta^{i}\right)\right) > 0.$$

Clearly, 给定 $\delta \in (0, 1)$ the last inequality $\sum_{n=3}^{\infty} \frac{1}{n!}(\lambda_1 t)^n \delta^2 (1-\delta)\left(\sum_{i=2}^{n-1} \delta^{n-1-i}\left(1-\delta^{i-1}\right)\left(1-\delta^{i}\right)\right) > 0, \forall t > 0$ holds given that $\delta \in (0, 1)$, which indicates $R_S^{1,2}(t) > R_S^{2,1}(t), \forall t > 0$.

Combining the results for $\lambda_2 \neq \delta\lambda_1$ and $\lambda_2 = \delta\lambda_1$, we prove the Theorem 6.2.

With Theorem 6.2, it is clear that using the weaker component as the primary component in the 1-out-of-2 warm standby system would lead to a better system performance in terms of the system reliability. Note that $R_S^{1,2}(t) > R_S^{2,1}(t)$ can directly leads to $E\left[T_S^{1,2}\right] > E\left[T_S^{2,1}\right]$, but the converse does not necessarily hold. So Theorem 6.2 is a stronger result than Theorem 6.1 for the linear AFTM case.

Example 6.2. Assume $\gamma(t) = 0.5t$. Consider a 1-out-of-2 warm standby system. with two exponentially distributed components A_1 and A_2, where the failure rates $\lambda_1 = 0.1$ and $\lambda_2 = 0.01$, respectively. Figure 6.1 shows the system reliability functions with different component orders. Apparently, the system reliability function with structure 1,2 is always better than that with structure 2,1, indicating that the weaker component should be used first to achieve a higher system reliability.

6.3 Optimal Component Working Order in General 1-Out-Of-*N* System

This section discusses the optimal component working order in general 1-out-of-n systems. Similar as for the 1-out-of-2 system, we consider the expected system lifetime and the system reliability as two reliability indicators.

6.3.1 Optimal Working Component Order Maximizing the Expected System Lifetime

Suppose that two warm standby components A_{n-1} and A_n are to be added into a 1-out-of-$(n-2)$ standby system to further enhance the system reliability. Then there would be an optimal allocation problem: to achieve a higher expected syste lifetime, how do we arrange the order of these two components? Theorem 6.3 is proposed to solve this problem.

Theorem 6.3. Suppose $\gamma(t)$ satisfies $\gamma'(t) = d\gamma(t)/dt \in (0,1), \gamma(0) = 0, \lim_{t\to+\infty}\gamma(t) = +\infty$. For a fixed 1-out-of-($n$-2) system, we add two more warm standby components into the system to form a 1-out-of-n standby system. If the failure rate functions of the two added components satisfy $\lambda_{n-1}(t) > \lambda_n(t), \forall t > 0$, then we have

$$E\left[T_S^{1,\ldots,n-2,n-1,n}\right] > E\left[T_S^{1,\ldots,n-2,n,n-1}\right],$$

i.e., a larger expected system lifetime can be achieved if the component with higher failure rate would be first used.

Proof. With the recursive relation in (6.8), the fixed 1-out-of-(n-2) system can be regarded as a single component. Theorem 6.3 holds as long as it holds for $n = 3$, i.e., a 1-out-of-3 warm standby system. So we only consider a 1-out-of-3 warm standby system.

With (6.7), and (6.9), we have

$$\begin{aligned} E\left[T_S^{1,2,3}\right] &= E\left[T_S^{1,2}\right] + E[T_3] - \int_0^{+\infty} R_3(w) R_S^{1,2}\left(\gamma^{-1}(w)\right) dw \\ &= E\left[T_S^{1,2}\right] + E[T_3] - \int_0^{+\infty} R_3(\gamma(u)) R_S^{1,2}(u) d\gamma(u) \\ &= E[T_1] + E[T_2] - \int_0^{+\infty} R_2(w) R_1\left(\gamma^{-1}(w)\right) dw + E[T_3] \\ &\quad - \int_0^{+\infty} R_3(\gamma(u)) \left(R_1(u) + \int_0^u R_2(u - \tau + \gamma(\tau)) f_1(\tau) d\tau \right) d\gamma(u) \\ &= E[T_1] + E[T_2] + E[T_3] - \int_0^{+\infty} R_2(\gamma(u)) R_1(u) d\gamma(u) \\ &\quad - \int_0^{+\infty} R_3(\gamma(u)) R_1(u) d\gamma(u) - \int_0^{+\infty} R_3(\gamma(u)) \\ &\quad \left(\int_0^u R_2(u - \tau + \gamma(\tau)) f_1(\tau) d\tau \right) d\gamma(u). \end{aligned} \tag{6.14}$$

Similarly,

$$E\left[T_S^{1,3,2}\right] = E[T_1] + E[T_2] + E[T_3] - \int_0^{+\infty} R_3(\gamma(u))R_1(u)d\gamma(u) - \int_0^{+\infty} R_2(\gamma(u))R_1(u)d\gamma(u) - \int_0^{+\infty} R_2(\gamma(u)) \left(\int_0^u R_3(u-\tau+\gamma(\tau))f_1(\tau)d\tau\right)d\gamma(u). \tag{6.15}$$

Comparing (6.14) and (6.15), it is sufficient to prove $E\left[T_S^{1,2,3}\right] > E\left[T_S^{1,3,2}\right]$ if

$$R_3(\gamma(u))R_2(u-\tau+\gamma(\tau))f_1(\tau) < R_2(\gamma(u))R_3(u-\tau+\gamma(\tau))f_1(\tau),$$

i.e.,

$$R_3(\gamma(u))R_2(u-\tau+\gamma(\tau)) < R_2(\gamma(u))R_3(u-\tau+\gamma(\tau))$$

holds for $0 < \tau < u$.
With the basic relationship $R(t) = \exp\left\{-\int_0^t \lambda(\tau)d\tau\right\}$, we have

$$\begin{aligned}
&R_3(\gamma(u))R_2(u-\tau+\gamma(\tau)) < R_2(\gamma(u))R_3(u-\tau+\gamma(\tau))\\
\Leftrightarrow\ &\frac{R_2(u-\tau+\gamma(\tau))}{R_2(\gamma(u))} < \frac{R_3(u-\tau+\gamma(\tau))}{R_3(\gamma(u))}\\
\Leftrightarrow\ &\frac{\exp\left\{-\int_0^{u-\tau+\gamma(\tau)}\lambda_2(t)dt\right\}}{\exp\left\{-\int_0^{\gamma(u)}\lambda_2(t)dt\right\}} < \frac{\exp\left\{-\int_0^{u-\tau+\gamma(\tau)}\lambda_3(t)dt\right\}}{\exp\left\{-\int_0^{\gamma(u)}\lambda_3(t)dt\right\}}\\
\Leftrightarrow\ &\exp\left\{-\int_{\gamma(u)}^{u-\tau+\gamma(\tau)}\lambda_2(t)dt\right\} < \exp\left\{-\int_{\gamma(u)}^{u-\tau+\gamma(\tau)}\lambda_3(t)dt\right\}\\
\Leftrightarrow\ &\int_{\gamma(u)}^{u-\tau+\gamma(\tau)}\lambda_2(t)dt > \int_{\gamma(u)}^{u-\tau+\gamma(\tau)}\lambda_3(t)dt.
\end{aligned} \tag{6.16}$$

With the assumption$d\gamma(t)/dt \in (0, 1)$, it is clear that function $g(t) = t - \gamma(t)$ is strictly increasing, which implies $\tau - \gamma(\tau) < u - \gamma(u)$, i.e., $\gamma(u) < u - \tau + \gamma(\tau)$ for$0 < \tau < u$. Therefore, $\int_{\gamma(u)}^{u-\tau+\gamma(\tau)}\lambda_2(t)dt > \int_{\gamma(u)}^{u-\tau+\gamma(\tau)}\lambda_3(t)dt$ holds given$\lambda_2(t) > \lambda_3(t)$, which completes the proof.

Here, $d\gamma(t)/dt \in (0, 1)$ has concrete physical meaning: the consumption of the virtual age in the warm standby state is slower than that in the operational state. Theorem 6.3 implies that, if we have to add two different warm standby components into the original system to improve the system performance, the weaker component should be used first to achieve a larger expected system lifetime. Obviously, the 1-out-of-3 warm standby system is a special case of Theorem 6.3. For a 1-out-of-3 warm standby system, if the primary component is fixed, then the weaker component with higher failure rate should be used as the first standby component to achieve a higher expected system lifetime.

Example 6.3. Consider a 1-out-of-3 warm standby system. Suppose the primary component A_1 with reliability function $R_1(t) = \exp\{-(t/100)^2\}$ is predetermined. The failure rate functions of two standby components A_2 and A_3 are $\lambda_2(t) = 0.01 + 2 \times 10^{-4}t$ and $\lambda_3(t) = 2 \times 10^{-4}t$, respectively. Suppose $\gamma(t) = 0.5t$. Then we have $E\left[T_S^{1,2,3}\right] = 166.162$ and $E\left[T_S^{1,3,2}\right] = 164.579$. It is clear that we should use the weaker component first to achieve a higher expected system lifetime.

6.3.2 *Optimal Working Component Order Maximizing the System Reliability*

Still consider the scenario in Theorem 6.3, i.e., two warm standby components A_{n-1} and A_n are added into a 1-out-of-(n-2) standby system to consititute a 1-out-of-n standby system. Suppose the AFTM is linear, and the liftimes of the added components are exponentially distributed. Then we have the following theorem.

Theorem 6.4. Consider the linear AFTM $\gamma(t) = \gamma t, \gamma \in (0, 1)$. For a fixed 1-out-of-($n$-2) system, consider adding two more exponentially distributed warm standby components into the system to form a 1-out-of-n standby system. Suppose the failure rates of the two warm standby components satisfy $\lambda_{n-1} > \lambda_n$. Then we have

$$R_S^{1,\ldots,(n-2),(n-1),n}(t) > R_S^{1,\ldots,(n-2),n,(n-1)}(t), \forall t > 0.$$

Proof. With the recursive formula in (6.8), regarding the fixed 1-out-of-(n-2) system as a single components, Theorem 6.4 holds as long as it holds for $n = 3$, i.e., the 1-out-of-3 warm standby system. So we only consider a 1-out-of-3 warm standby system. Denote

$$H_S^{1,2,3}(t) = \int_0^t R_3(t-\tau+\gamma(\tau))\left(\int_0^\tau f_2(\tau-u+\gamma(u))f_1(u)du - R_2(\gamma(\tau))f_1(\tau)\right)d\tau,$$

$$H_S^{1,3,2}(t) = \int_0^t R_2(t-\tau+\gamma(\tau))\left(\int_0^\tau f_3(\tau-u+\gamma(u))f_1(u)du - R_3(\gamma(\tau))f_1(\tau)\right)d\tau.$$

According to (6.7), $R_S^{1,2,3}(t) > R_S^{1,3,2}(t)$ holds if $H_S^{1,2,3}(t) > H_S^{1,3,2}(t)$. For the 1-out-of-3 warm standby system with exponential standby components, we have

$$H_S^{1,\,2,\,3}(t) = \int_0^t e^{-\lambda_3(t-\tau+\gamma\tau)}\left(\int_0^\tau \lambda_2 e^{-\lambda_2(\tau-u+\gamma u)} f_1(u)du - e^{-\lambda_2\gamma\tau} f_1(\tau)\right)d\tau$$

$$\overset{\delta=1-\gamma}{\Leftrightarrow} e^{-\lambda_2 t}\int_0^t e^{\lambda_2\delta\tau}\left(\int_0^\tau \lambda_3 e^{-\lambda_3\tau+\lambda_3\delta u} f_1(u)du - e^{-\lambda_3\gamma\tau} f_1(\tau)\right)d\tau$$

$$= \int_0^t \frac{\lambda_2 e^{-\lambda_3\gamma(t-\tau)-\lambda_2(t-\tau)} - \lambda_3\delta e^{-\lambda_3(t-\tau)}}{\lambda_3\delta - \lambda_2} e^{-\lambda_3\gamma\tau-\lambda_2\gamma\tau} f_1(\tau)d\tau.$$

Similarly,

$$H_S^{1,\,3,\,2}(t) = \int_0^t e^{-\lambda_2(t-\tau+\gamma\tau)}\left(\int_0^\tau \lambda_3 e^{-\lambda_3(\tau-u+\gamma u)} f_1(u)du - e^{-\lambda_3\gamma\tau} f_1(\tau)\right)d\tau$$

$$\overset{\delta=1-\gamma}{\Leftrightarrow} e^{-\lambda_2 t}\int_0^t e^{\lambda_2\delta\tau}\left(\int_0^\tau \lambda_3 e^{-\lambda_3\tau+\lambda_3\delta u} f_1(u)du - e^{-\lambda_3\gamma\tau} f_1(\tau)\right)d\tau$$

$$= \begin{cases} \int_0^t \frac{\lambda_3 e^{-\lambda_2\gamma(t-\tau)-\lambda_3(t-\tau)} - \lambda_2\delta e^{-\lambda_2(t-\tau)}}{\lambda_2\delta-\lambda_3} e^{-\lambda_2\gamma\tau-\lambda_3\gamma\tau} f_1(\tau)d\tau, & \lambda_3 \neq \lambda_2\delta \\ \int_0^t e^{-\lambda_2(t-\tau)}(\lambda_3(t-\tau)-1)e^{-\lambda_2\gamma\tau-\lambda_3\gamma\tau} f_1(\tau)d\tau, & \lambda_3 = \lambda_2\delta \end{cases}.$$

We consider the two cases $\lambda_3 \neq \lambda_2\delta$ and $\lambda_3 = \lambda_2\delta$.

(1) $\lambda_3 \neq \lambda_2\delta$

It is sufficient to prove the following inequality for $0 < \tau < t$ to prove $H_S^{1,2,3}(t) > H_S^{1,3,2}(t)$:

$$\frac{\lambda_2 e^{-\lambda_3\gamma(t-\tau)-\lambda_2(t-\tau)} - \lambda_3\delta e^{-\lambda_3(t-\tau)}}{\lambda_3\delta - \lambda_2} e^{-\lambda_3\gamma\tau-\lambda_2\gamma\tau} f_1(\tau)$$
$$> \frac{\lambda_3 e^{-\lambda_2\gamma(t-\tau)-\lambda_3(t-\tau)} - \lambda_2\delta e^{-\lambda_2(t-\tau)}}{\lambda_2\delta - \lambda_3} e^{-\lambda_2\gamma\tau-\lambda_3\gamma\tau} f_1(\tau)$$
$$\Leftrightarrow \frac{\lambda_2 e^{-\lambda_3\gamma(t-\tau)-\lambda_2(t-\tau)} - \lambda_3\delta e^{-\lambda_3(t-\tau)}}{\lambda_3\delta - \lambda_2} > \frac{\lambda_3 e^{-\lambda_2\gamma(t-\tau)-\lambda_3(t-\tau)} - \lambda_2\delta e^{-\lambda_2(t-\tau)}}{\lambda_2\delta - \lambda_3}.$$

Denote $u = t - \tau$. The above inequality is equivalent to the followings

$$\begin{aligned} &\frac{\lambda_2 e^{-\lambda_3\gamma u-\lambda_2 u} - \lambda_3\delta e^{-\lambda_3 u}}{\lambda_3\delta - \lambda_2} > \frac{\lambda_3 e^{-\lambda_2\gamma u-\lambda_3 u} - \lambda_2\delta e^{-\lambda_2 u}}{\lambda_2\delta - \lambda_3} \\ &\Leftrightarrow \frac{\delta\lambda_2^2\left(e^{\lambda_3 u} - e^{\lambda_3\delta u}\right) + \lambda_2\lambda_3\left(e^{\lambda_3\delta u} - \delta^2 e^{\lambda_3 u}\right)}{(\lambda_2 - \lambda_3\delta)(\lambda_3 - \lambda_2\delta)} \\ &< \frac{\lambda_3^2\delta\left(e^{\lambda_2 u} - e^{\lambda_2\delta u}\right) + \lambda_2\lambda_3\left(e^{\lambda_2\delta u} + \delta^2 e^{\lambda_2 u}\right)}{(\lambda_2 - \lambda_3\delta)(\lambda_3 - \lambda_2\delta)}. \end{aligned} \tag{6.17}$$

Comparing (6.17) and (6.12), we can see that (6.17) holds if and only if (6.12) holds. We have shown that (6.12) holds. Therefore, (6.17) holds, and $H_S^{1,2,3}(t) > H_S^{1,3,2}(t)$.

(2) $\lambda_3 = \lambda_2\delta$

To prove $H_S^{1,2,3}(t) > H_S^{1,3,2}(t)$, we only need the following inequality

$$
\begin{aligned}
&\frac{\lambda_2 e^{-\lambda_3\gamma(t-\tau)-\lambda_2(t-\tau)} - \lambda_3\delta e^{-\lambda_3(t-\tau)}}{\lambda_3\delta - \lambda_2} e^{-\lambda_3\gamma\tau-\lambda_2\gamma\tau} f_1(\tau) \\
&\quad > e^{-\lambda_2(t-\tau)}(\lambda_3(t-\tau)-1)e^{-\lambda_2\gamma\tau-\lambda_3\gamma\tau} f_1(\tau) \\
&\Leftrightarrow \frac{\lambda_3\delta e^{\lambda_2(t-\tau)} - \lambda_2 e^{\lambda_3\delta(t-\tau)}}{\lambda_2 - \lambda_3\delta} > e^{\lambda_3(t-\tau)}(\lambda_3(t-\tau)-1) \\
&\underset{u=t-\tau}{\overset{\lambda_3=\lambda_2\delta}{=}} \quad (1-\delta^2)e^{\delta\lambda_2 u} + \delta^2 e^{\lambda_2 u} - e^{\delta^2\lambda_2 u} > (1-\delta^2)\delta\lambda_2 u e^{\delta\lambda_2 u}.
\end{aligned} \tag{6.18}
$$

Comparing (6.18) and (6.13), we can see that (6.18) holds if and only if (6.13) holds. We have shown that (6.13) holds. Therefore, (6.18) holds and $H_S^{1,2,3}(t) > H_S^{1,3,2}(t)$.

Combining the results for $\lambda_2 \neq \delta\lambda_1$ and $\lambda_2 = \delta\lambda_1$, Theorem 6.4 is proved.

In Theorem 6.2, we prove that, for a 1-out-of-2 warm standby system with exponential components, using the weaker components as the primary component could lead to a higher system reliability. Theorem 6.4 implies that, for two additional exponentially distributed warm standby components, using the weaker component first could achieve higher system reliability. It is interesting to investigate, for a general 1-out-of-n warm standby system with exponential components, whether using the weaker component first would lead to a more reliable system. Theorem 6.5 is proposed to solve this problem.

Theorem 6.5. Consider the linear AFTM $\gamma(t) = \gamma t, \gamma \in (0, 1)$. For a 1-out-of-$n$ warm standby system with exponential components, suppose the failure rates of the n components satisfy $\lambda_1 > \lambda_2 > \ldots > \lambda_n$. Then the system with structure "$1, 2, \ldots, n$" is the most reliable system, i.e.,$R_S^{1,2,\ldots,n}(t) > R_S^{i_1,i_2,\ldots,i_n}(t), \forall t > 0$ holds for any $(i_1, i_2, \ldots, i_n) \neq (1, 2, \ldots, n)$ where $i_1, i_2, \ldots, i_n$ can be any other permutation of $1, 2, \ldots, n$.

Theorem 3 in [3] is required to prove Theorem 6.5. For ease of the reference, we rewrite this theorem as Lemma 6.2 with a slight modification. For more details about this lemma, refer to [3].

Lemma 6.2. If $d\gamma(t)/dt \in (0, 1)$, then $R_S^{1',2}(t) > R_S^{1,2}(t), \forall t > 0$ holds given $R_{1'}(t) > R_1(t), \forall t > 0$, where $R_{1'}(t)$ and $R_1(t)$ are the reliability functions of two components $A_{1'}$ and A_1.

The proof for Theorem 6.5 could be done using the idea of a bubble sort. Suppose "$i_1, i_2, \ldots, i_n$" is a permutation of "$1, 2, \ldots, n$", and the system reliability of the 1-out-of-n warm standby system with structure $i_1, i_2, \ldots, i_n$ is $R_S^{i_1,i_2,\ldots,i_n}(t)$.

If $i_1 > i_2$, then with Theorem 6.2 we know that

$$R_S^{i_2,i_1}(t) > R_S^{i_1,i_2}(t),$$

i.e., the 1-out-of-2 warm standby system should use the component with higher failure rate as the primary component.

With Lemma 6.2 and recursive formula (6.8), we could prove that

$$R_S^{i_2,i_1,i_3}(t) > R_S^{i_1,i_2,i_3}(t).$$

Repeat applying Lemma 6.2 and the recursive formula, we have

$$R_S^{i_2,i_1,i_3,\ldots,i_n}(t) > R_S^{i_1,i_2,i_3,\ldots,i_n}(t).$$

If $i_k > i_{k+1} (1 < k < n)$, then according to Theorem 6.4 we have

$$R_S^{i_1,\ldots,i_{k-1},i_{k+1},i_k}(t) > R_S^{i_1,\ldots,i_{k-1},i_k,i_{k+1}}(t),$$

i.e., if we add two components to a 1-out-of-(k-1) warm standby system with component working order $i_1, \ldots, i_{k-1}$, the component with higher failure rate $A_{i_{k+1}}$ should be used first. Repeat applying Lemma 6.2 and the recursive formula (6.8), we have

$$R_S^{i_1,\ldots,i_{k-1},i_{k+1},i_k,i_{k+2},\ldots,i_n}(t) > R_S^{i_1,\ldots,i_{k-1},i_k,i_{k+1},i_{k+2},\ldots,i_n}(t).$$

So, for an arbitrary permutation of "$1, 2, \ldots, n$", as long as there exists two adjacent numbers that satisfy $i_k > i_{k+1}$ in "$i_1, i_2, \ldots, i_n$", we should swap them to achieve a higher system reliability. By repeatedly stepping through i_1 to i_n, we would have "$1, 2, \ldots, n$" in the end, which implies that the structure "$1, 2, \ldots, n$" could lead to the most reliable warm standby system.

Theorem 6.5 implies that, for a 1-out-of-n system with exponential components, we should arrange the positions of the components in descending order of their failure rates for achieving the highest system reliability.

Example 6.4. Consider a 1-out-of-3 warm standby system. Suppose three components A_1, A_2 and A_3 are exponentially distributed with failure rates $\lambda_1 = 0.1$, $\lambda_2 = 0.01$ and $\lambda_3 = 0.001$, respectively. Suppose $\gamma(t) = 0.5t$. The system reliability functions with different component orders are illustrated in Fig. 6.2. It can be seen that the system with structure "1, 2, 3" is the most reliable, which again implies that we should use the weaker component first to achieve a higher system reliability for the warm standby system.

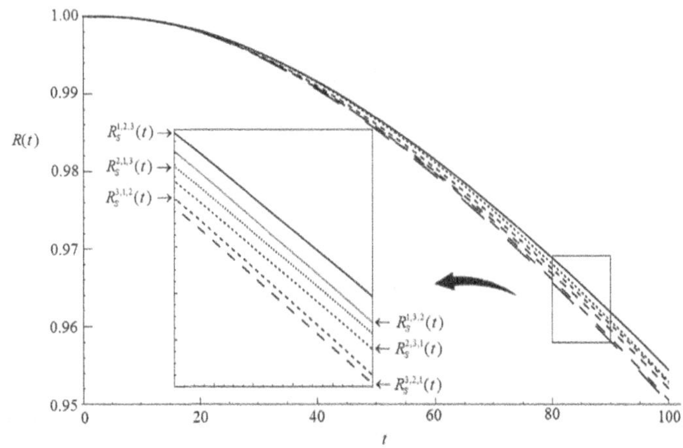

Figure. 6.2 System reliability functions with different component orders

6.4 Conclusion

In this chapter, the optimal component working order for the 1-out-of-n warm standby system is investigated. Higher system reliability could be achieved by properly arranging the component order in the system. It is proved that, for a 1-out-of-2 warm standby system, using the weaker component as the active one could achieve a larger expected system lifetime. It is further proved that, for a fixed 1-out-of-(n-2) system, if we add another two warm standby components into the system to form a 1-out-of-n standby system, then using the weaker component as the (n-1)th standby component would lead to a larger expected system lifetime. For the special case of linear AFTM and exponential components, it is proved that, for a 1-out-of-n warm standby system, arranging the components in descending order of their failure rates leads to the most reliable system.

References

1. Kay R, Kinnersley N. On the use of the accelerated failure time model as an alternative to the proportional hazards model in the treatment of time to event data: A case study in influenza [J]. Drug Information Journal, 2002, 36(3): 571-9.
2. Finkelstein M. On statistical and information-based virtual age of degrading systems [J]. Reliability Engineering & System Safety, 2007, 92(5): 676-81
3. Cha J H, Mi J, Yun W Mo Ydelling a general standby system and evaluation of its performance [J]. Applied Stochastic Models in Business and Industry, 2008, 24(2): 159-69

Chapter 7
Reliability Evaluation for Demand-Based Warm Standby Systems Considering Degradation Process

Warm standby redundancy is a fault-tolerant technique balancing the low economical efficiency of hot standby and the long recovery time of cold standby. Motivated by practical engineering systems, a general demand-based warm standby system (DB-WSS) considering component degradation process is studied. A series of intermediate states exists between perfect functionality and complete failure because of degradation processes. A lot of existing analytical reliability assessment techniques are focused on conventional binary-state models or exponential state transition distributions for a system or its components. A novel reliability evaluation approach based on the multi-state decision diagram (MSDD) for DB-WSS is proposed. The proposed technique can handle arbitrary distributions of degradation processes for multi-state components or systems. Moreover, considering the imperfect switch of the warm standby component, the start failure probability is taken into account in the warm standby system (WSS). Numerical studies are given to illustrate the proposed approach.

7.1 Introduction

Standby redundancy is widely implemented in many critical engineering systems such as electric power systems, computing systems and flight control systems for improving corresponding reliability [1, 2]. Standby redundancy can be further classified into hot standby, warm standby and cold standby [3]. Hot standby is a redundancy technique in which standby components operate simultaneously with primary components [4]. Once the primary components fail or the capacity of primary components cannot satisfy system demand, and the hot standby immediately takes them over. The redundant components in hot standby usually have the same failure rates as those primary components in operations [5]. Cold standby involves a backup system for another primary system [6]. The cold standby is called upon only after failure or insufficient capacity of the primary system. It may take a few hours or more to

R. Peng et al., *Reliability Modelling and Optimization of Warm Standby Systems*,
https://doi.org/10.1007/978-981-16-1792-8_7

start up the cold standby components. The cold standby components do not fail until switched on [5]. Compared with the hot standby, cold standby takes lower cost but longer recovery time while the hot standby may cause more energy consumption [7]. To balance the system economical efficiency and the recovery time, warm standby rises. Warm standby indicates that one system operates in the background of the primary system [8]. Warm standby components have different state transition rates or even different state transition distributions before and after they become operating components [1]. For example, in a power system, the supply of electric power is supposed to balance the system demand over a short term. To allow for demand real-time balancing, generating units may be kept running with the generators switched on. Therefore, when peaks of demand occur, the generators can be immediately switched on to balance the demand. Being in the state ready to operate is known as hot standby. If the electric power deficiency is too large to be balanced or the system is in contingency states, some generating units, whose turbo-alternators are shut down while the boilers are left in hot conditions, can start up and synchronize with the system with a very short lead time [9]. This kind of generating units can be defined as one type of warm standby. Cold standby is usually utilized in long-term planning and design for power systems due to the significant amount of starting time for steam turbines.

However, the system may fail if the standby components suffer a start-up failure [10] or imperfect switching [11]. Therefore, the start failure probabilities of the warm standby components are supposed to be taken into account.

A demand-based system indicates that the number of online components depends on the system demand [12, 13]. When the summation of the components' capacities is sufficient to cover the demand requirement, these components are in operating states, and the other components are in warm standby states. This kind of system is known as the demand-based warm standby system (DB-WSS). This includes power systems in which power generation is intended to meet the time-varying power demand in operating time [14].

The classical technique for system reliability analysis is the binary-state model. Reliability analysis on traditional binary-state models is only valid for two possible states and for its components: perfect functionality and complete failure. Binary-state reliability analysis technique is the classical reliability analysis method and has been intensively applied to engineering problems [15].

However, the degradation process often occurs in systems until they fail completely. Therefore, a series of intermediate states exists between perfect functioning and complete down, and the concept of the multi-state system (MSS) was introduced [16]. The multi-state model for degradation processes is extensively applied because it practically fits to system aging processes in real-life situations where there is a range of levels from perfect functionality to complete failure [17]. To model the dynamic process of degradation, stochastic models such as Markov [18, 19] or semi-Markov [20, 21] models have been utilized [22]. In Markov models, the transition time distributions for degradation processes follow exponential distributions, which means that the degradation process is memoryless. In semi-Markov models, the transition time distributions for degradation processes may follow arbitrary type of distributions.

Two major methods to evaluate MSS reliability are Monte Carlo simulation techniques [23–25] and analytical methods [26]. The simulation methods are flexible and can be easily modified to adapt to different situations. However, the simulation techniques require much longer computing time for systems with a great number of components and only provide approximate results. Analytical techniques are efficient for assessing system reliability indices. The analytical approaches are usually faster than simulation techniques [26]. However, there exist limitations for different analytical techniques. With the increase in the number of components for the system and the number of states for components, the analytical modeling for reliability evaluation may become complex [25].

The decision diagram technique [12] is one of the analytical methods. Reference [27] exploits three forms of decision diagrams, including binary decision diagrams (BDD), logarithmically encoded BDD and multi-valued decision diagrams to evaluate MSS reliability. This type of analytic method, however, may need a pre-processing step to obtain probability values for each state of components with state transition processes and may not be applied to redundant systems such as WSS. Reference [28] proposed a BDD-based reliability evaluation method for WSS without specifically obtaining probability values for each state of components. Similarly, the multi-valued decision diagram is used in reliability analysis of the phased-mission system [12, 29, 30]. However, these two approaches mainly focus on reliability assessment of systems with binary-state components, without taking multi-state components into account.

The combined stochastic process [31] method and universal generating function (UGF) technique [32–36] or Lz transform approach [37, 38] has been widely used in reliability evaluation for MSS [39, 40]. This combined method provides a comprehensive approach for the system state enumeration that can substitute complicated combinational algorithms and reduce the computational complexities [32]. References [33, 40] have respectively proposed a technique based on the combination of UGF or Lz transform approach and Markov process methods to handle reliability evaluation for MSS. However, the conventional Markov process is oriented to state space and deals with only exponential distributions [41]. To the best of our knowledge, the UGF method or Lz transform approach usually does not consider the chronological characteristics of components in operational phases for WSS and therefore they may not be directly applied to WSS.

Motivated by practical engineering systems such as power systems mentioned above, we have made contributions by analyzing general DB-WSS considering degradation process that practically fits the aging process in real life situations compared with the classical binary-state model. Moreover, a multi-state decision diagram (MSDD) method has been proposed for reliability evaluation that is applicable and flexible with arbitrary types of state transition distributions (not only restricted to exponential distributions) for DB-WSS. The proposed MSDD-based technique is an effective analytic approach for reliability evaluation with accurate results. The MSDD based approach is extended from the BDD method in reference [28]. The proposed method is applicable to systems with multi-state components considering degradation process without the limitation of reliability analysis only for binary-state components in [28]. Consequently, the application scope of the proposed technique is expanded. Moreover, the start failure probabilities are taken into account in the WSS considering the imperfect switches of warm standby components.

7.2 System Descriptions and Assumptions

The system description and several assumptions are explained as follows.

(1) The DB-WSS with parallel structure is composed of N components, which are not necessarily identical.

(2) The number of the online components depends on the system demand. If the total capacity of former k components ($A_1, A_2, \ldots, A_k$) satisfies the predetermined demand, the remaining ($N-k$) components ($A_{k+1}, A_{k+2}, \ldots, A_N$) are in the warm standby state. A_i^o indicates that component A_i is in operating state and A_i^w denotes that component A_i is in the warm standby state. The structure of the DB-WSS is shown in Fig. 7.1.

(3) The degradation processes occur to both online and warm standby components. Because of degradation processes, each component A_i can be in M_i+1 possible degradation states ($D_0, D_1, \ldots, D_{M_i}$) with corresponding decreasing capacities. Each state D_j of component A_i is assumed to have a capacity $C_{D_j}^{A_i}$.D_0 is the initial state with fully powered up capacity $C_{D_0}^{A_i}$. The degradation transition process is between two adjacent states which is from the state with larger capacity to the state with smaller capacity, i.e., the degradation process can only transfer from D_j to $D_{j+1}, j = 0, \ldots, M_i - 1$. Figure 7.2 illustrates the degradation processes of a multi-state component A_i.

(4) When a degradation process occurs to the online component, one or more warm standby components will be powered up to ensure that the total capacity

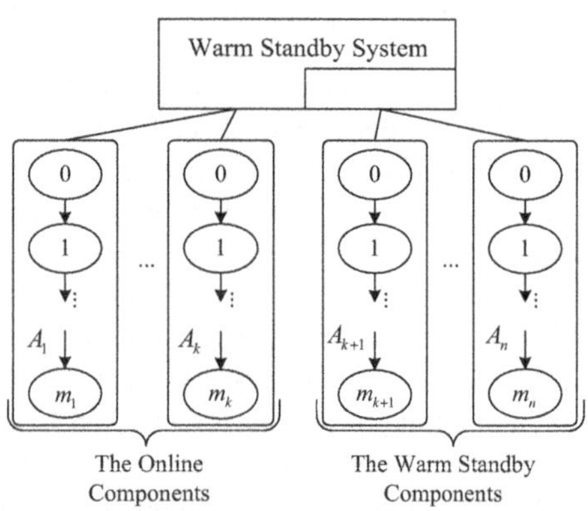

Fig. 7.1 The structure of DB-WSS

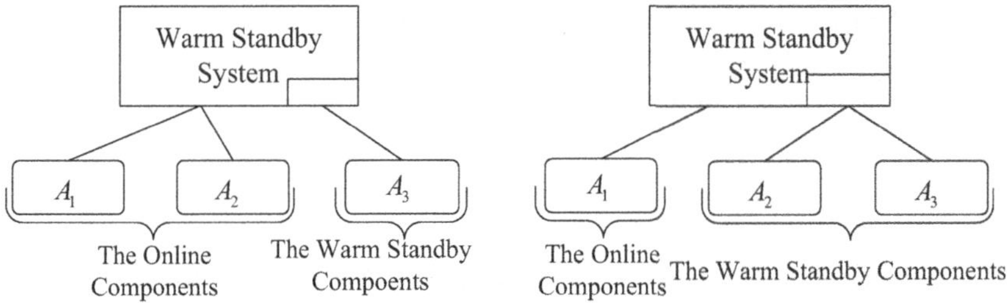

(a) The system structure when the system requirement is 7 (b) The system structure when the system requirement is 5

Fig. 7.2 The degradation process of a multi-state component A_i

of functioning components is not smaller than the required system demand rd. Clearly, when $\sum_{i=1}^{k} C_{D_j}^{A_i} < rd$, component A_{k+1} converts from the warm standby state to the operation state.

(5) The activation sequence of warm standby components is in a predetermined order. It is assumed that the discipline of starting warm standby components depends on the index of the warm standby components [13]. Specifically, the activation of warm standby component A_i occurs before the starting of component A_j for $k+1 \leq i < j \leq N$.

(6) The transition time distribution for the degradation process of state D_j for online component A_i is represented as $F_j^{A_i}(t)$, which indicates the transition of primary component A_i from state D_j to D_{j+1}. To account for the time-dependent degradation process of components in the system, different transition time distributions for degradation process are utilized before and after the warm standby components are fully powered up. Specifically, state D_j of warm standby component A_i^w is assumed to follow the transition time distribution $F_j^{A_i^w}(t)$ for degradation process at a warm standby state from state D_j to D_{j+1}.

(7) Considering systems which are impractical for performing maintenance [29], it is assumed that both the system and its components are non-repairable.

As an illustration, a DB-WSS with three components A_1,A_2 and A_3 is considered. The capacities of each degradation state for three components are listed in Table 7.1 Different system demand requirements are utilized to illustrate the demand-based

Table 7.1 The capacities of each degradation state for three components

Component/State	D_0	D_1	D_2	D_3
A_1	5	4	2	0
A_2	5	3	2	0
A_3	5	4	1	0

system. When the demand requirement is 7, components A_1 and A_2 are online components while component A_3 is in a warm standby state. When the system demand is 5, component A_1 is an operating component and components A_2,A_3 are warm standby components.

7.3 Reliability Evaluation Utilizing the MSDD Technique

7.3.1 The Construction of System MSDD

According to the system description and assumptions in Sect. 7.2, three aspects are supposed to be considered when constructing MSDD considering the degradation process. First, since the warm standby components may degrade either in a warm standby state or an operating state, the two different conditions should be dealt with via distinct modes. Second, since the warm standby components suffer the start failure probabilities, the number of activated standby components in the system should be incorporated in the MSDD illustration. Third, the capacity of each component should be reflected in the MSDD representation to determine the state of the system. Considering these three aspects, the traditional BDD should be extended to a MSDD.

Before the construction of the system level MSDD with degradation process, the illustration for MSDD about a single degradation should be constructed first. In the system operation period, a degradation process may occur from state D_j to D_{j+1} for a single component A_i, or there is no degradation for component A_i. The MSDD representation for each degradation process is built in an iterative method.

First, a system level MSDD representation of the first degradation is constructed. For the first time, its MSDD representation has $(N+1)$ branches. The i th $(1 \le i \le N)$ branch indicates that a degradation process occurs to component A_i, and the $(N+1)$ th branch denotes that there is no degradation occurring. The terminal value of a branch is expressed by a ternary $\{u_s^l(d, \varepsilon), C_s^l, q_s^l\}$. $u_s^l(d, \varepsilon)$ is a vector denoting the state of each component, where d indicates the degradation state of a component and ε indicates whether a component is in an online state or a warm standby state.

$$\varepsilon = \begin{cases} \alpha, & \text{the component is in online state} \\ \beta, & \text{the component is in warm standby state} \end{cases}$$

C_s^l denotes the available capacity of the system and $q_s^l = \left(q_{as}^l, q_{fs}^l\right)$ is a vector indicating the number of activated warm standby components and the number of failed components after the s th degradation for the l th branch. Before the first time, the available capacity of the system is $C_0^l = \sum_{i=1}^{N} C_{D_0}^{A_i}$ and the number of activated warm standby components and failed components is $q_0^l = (0, 0)$.

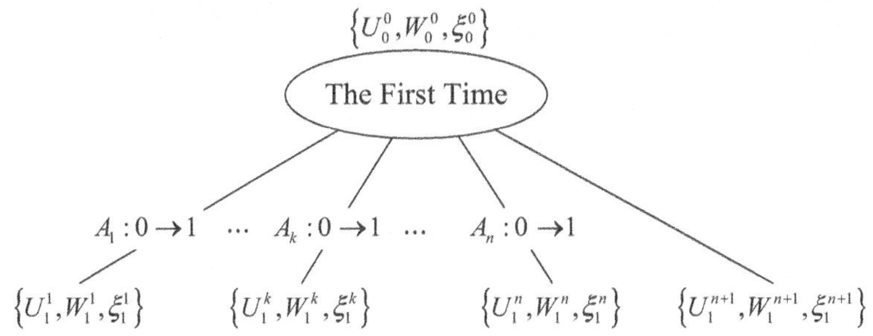

Fig. 7.3 The MSDD representation for the first time

Clearly, for the first time, the possible degradation process may occur to an arbitrary component from state D_0 to D_1, or there is no degradation at all. The MSDD representation for the first time is shown in Fig. 7.3. (It is assumed that there is no complete failure of a component after the first time.)

The system available capacity after the first time is

$$C_1^l = \begin{cases} C_{D_1}^{A_l} + \sum_{i \neq l} C_{D_0}^{A_i}, l \neq N+1 \\ \sum_{i=1}^{N} C_{D_0}^{A_i}, l = N+1 \end{cases}. \tag{7.1}$$

The number of newly activated standby components after the first time depends on the system demand and the degradation states of the online components. If the sum capacity of online components satisfies system demand, there is no need to activate warm standby components. Vice versa, if the sum capacity of online components is smaller than system demand, one or more warm standby components will be powered up until the system demand is met.

The MSDD representation for the second time is modified from the MSDD of the first time. The procedures for the second time are constructed as the following explanations. First, each terminal node in the MSDD representation for the first time (Fig. 7.3) is assumed to have $\left(N+1-q_{f1}^l\right)$ branches, where the ith $\left(1 \leq i \leq N-q_{f1}^l\right)$ branch indicates the degradation process of a component and the rightmost branch denotes that there is no degradation process. The leftmost branch in Fig. 7.3 is utilized for instance, which indicates the degradation process from state D_0 to D_1 of component A_1 after the first degradation. For the second time, the first branch has degradation process from state D_1 to D_2 of component A_1, and the ith $\left(2 \leq i \leq N-q_{f1}^l\right)$ branch indicates the degradation process from state D_0 to D_1 of any other components. The $\left(N+1-q_{f1}^l\right)$th branch denotes that there is no degradation process for the second time. Figure 7.4 illustrates the MSDD representation

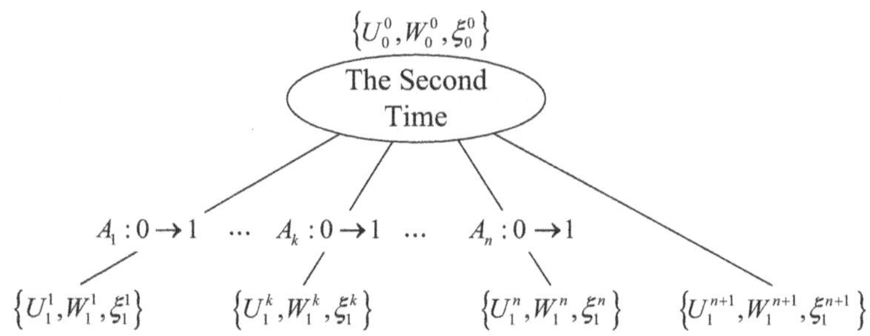

Fig. 7.4 The MSDD representation for the second time, given degradation process from state D_0 to D_1 of component A_1 as the first degradation

for the second time, given the degradation process from state D_0 to D_1 of component A_1 for the first degradation (if no component completely fails after the first time). Second, the new terminal value should be updated. The regenerated available capacity of the system after the second time is

$$C_2^l = \begin{cases} C_{D_2}^{A_1} + \sum_{i=2}^{N} C_{D_0}^{A_i}, l \leq 1 \\ C_{D_1}^{A_1} + C_{D_1}^{A_l} + \sum_{i \neq 1,l}^{N} C_{D_0}^{A_i}, 2 \leq l \leq N - q_{f1}^l \\ C_{D_1}^{A_1} + \sum_{i=2}^{N} C_{D_0}^{A_i}, l = N + 1 - q_{f1}^l \end{cases} \tag{7.2}$$

Similarly, the MSDD representation for the s th ($3 \leq s \leq \sum_{i=1}^{N} M_i$) degradation can be developed from the MSDD of the ($s - 1$) th degradation utilizing the above procedures.

Based on the MSDD of each individual degradation process, the system MSDD can be constructed in an iterative way. The overall construction of the system MSDD is illustrated in Fig. 7.5.

(1) Construct the MSDD representation of the first possible degradation process.
(2) Create the MSDD representation for the s th degradation based on the MSDD representation of the ($s - 1$) th degradation and update the corresponding terminal value of the system MSDD.
(3) If the available capacity is insufficient to meet the system demand or no more degradation can occur, the procedure is stopped and the system MSDD is obtained. That is, the construction of MSDD is stopped after the mth degradation as the formulation in Eq. (7.3).

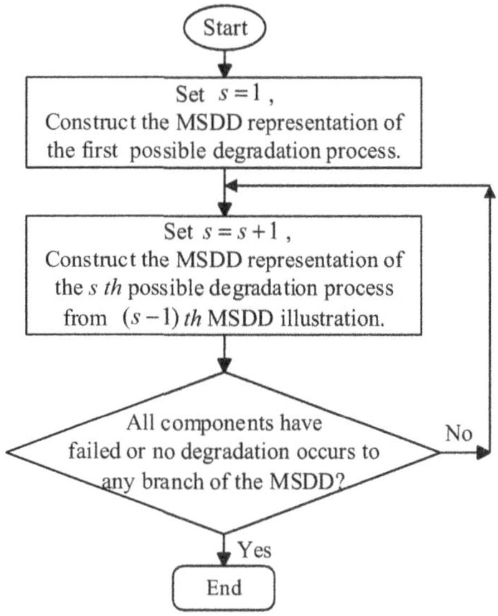

Fig. 7.5 The flowchart for the construction of MSDD

$$m = \max s,\ (C_s^l \geq rd,\ 1 \leq s \leq \sum_{i=1}^{N} M_i)\ or\ m = \sum_{i=1}^{N} M_i \tag{7.3}$$

The MSDD representation for each degradation process and the system MSDD are shown in Fig. 7.6. For simplification, the degradation state D_j is denoted by state j. The MSDD representation of the DB-WSS for the first time is shown in Fig. 7.6a. Figure 7.6b illustrates the MSDD representation of the second possible degradation process assuming that the first degradation process occurs to the online component A_1. Figure 7.6c describes the MSDD representation of the third possible degradation process on the hypothesis of component A_1 degrading twice.

To demonstrate the proposed MSDD construction, the example mentioned in Sect. 7.2 is utilized. Figure 7.7 demonstrates the MSDD construction for the provided example in Sect. 7.2. For simplification, the states of each component are used as terminal value in the MSDD representation in Fig. 7.7. For example, the node (1,0,0) in the first layer in Fig. 7.7a indicates that component A_1 is in state D_1, while components A_2 and A_3 are in state D_0. It should be noted that some subtrees can be merged if the parent nodes of these subtrees indicate the identical systems' states. For example, as shown in Fig. 7.7a, the two nodes (1,1,1) and their subtrees can be merged, correspondingly. In Fig. 7.7b, the node (1,1,0) and its subtrees are the same as the node (1,1,0) and the corresponding subtrees in Fig. 7.7a as well. Moreover, the node (1,2,0) in Fig. 7.7b indicates the same system state as node (1,2,0) in Fig. 7.7a. The node (1,1,1) and its subtrees in Fig. 7.7b are the same as those in Fig. 7.7a. The two nodes (0,2,1) in Fig. 7.7b denote identical system states and can be merged as well. Furthermore, in Fig. 7.7c, the node (1,0,1) and its subtrees are the same as node (1,0,1) and

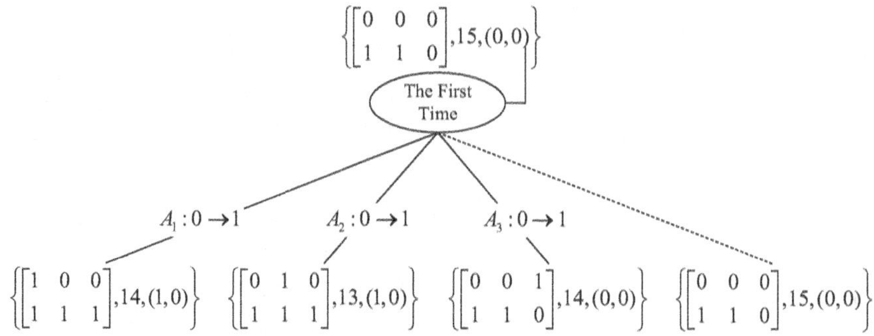

(a) The MSDD representation of the first time

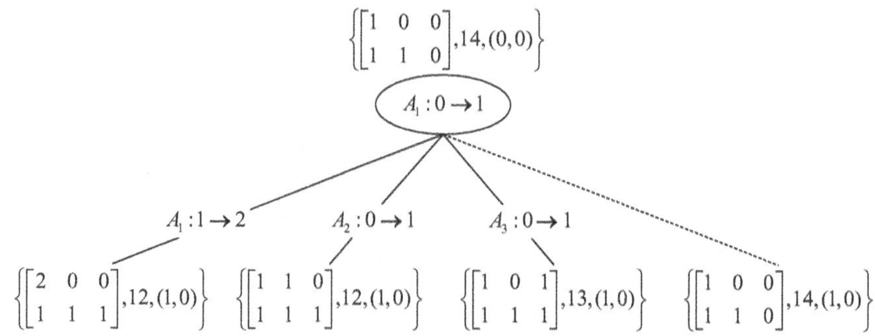

(b) The MSDD representation of the second time for the leftmost branch in Figure 7-6 (a)

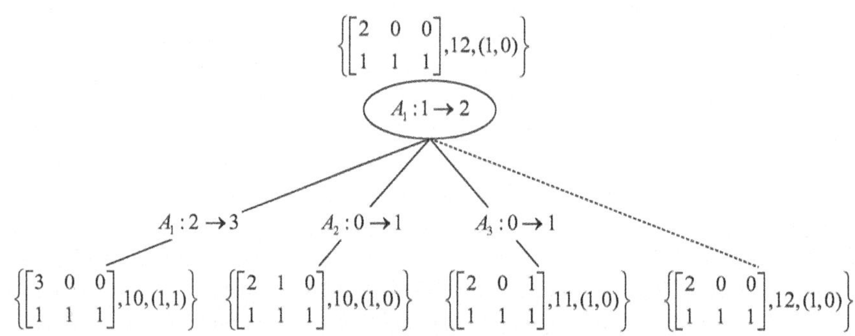

(c) The MSDD representation of the third time for the leftmost branch in Figure 7-6 (b)

Fig. 7.6 The MSDD construction of the proposed DB-WSS

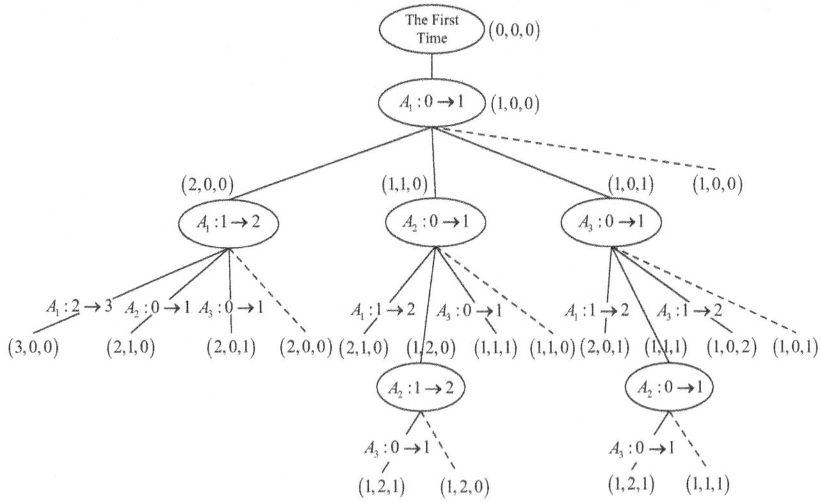

(a) The system MSDD representation for the first degradation from component A1

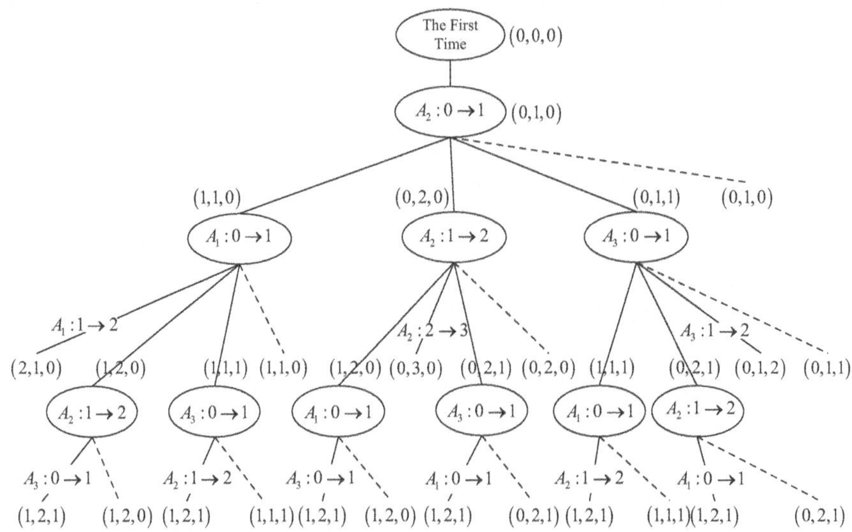

(b) The system MSDD representation for the first degradation from component A2

Fig. 7.7 The MSDD construction of the example for DB-WSS

the corresponding subtrees in Fig. 7.7a. The node (0,1,1) in Fig. 7.7c indicates the same system state as node (0,1,1) in Fig. 7.7b.

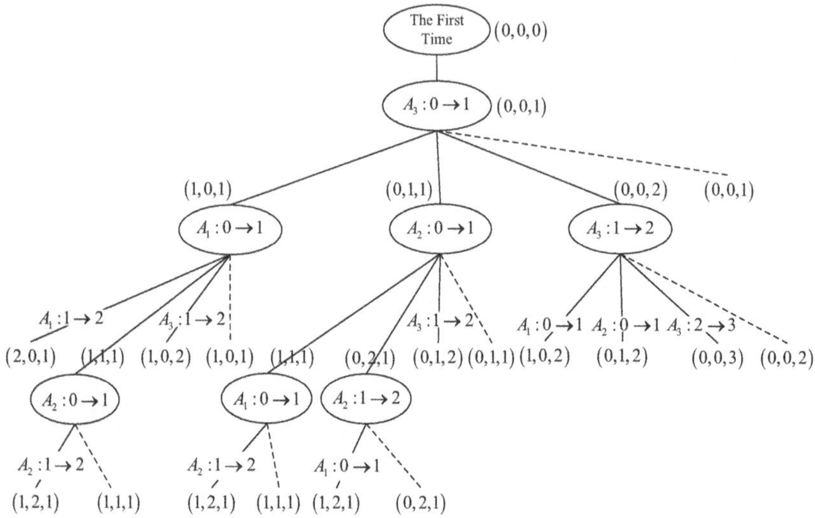

(c) The system MSDD representation for the first degradation from component A3

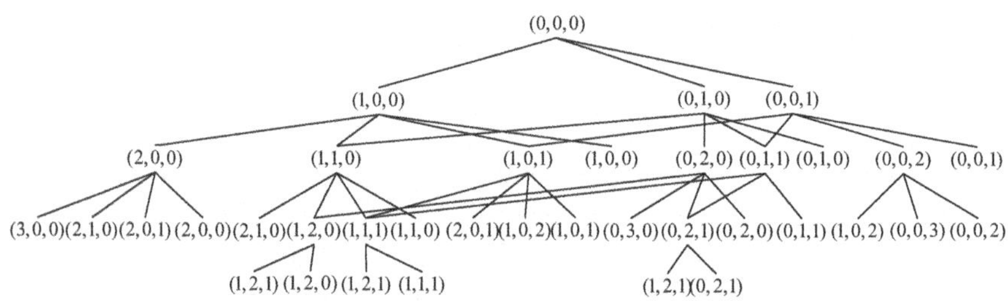

(d) The overall system MSDD representation for the example

Fig. 7.7 (continued)

7.3.2 System Reliability Evaluation Based on MSDD

The leftmost path in Fig. 7.7a "$(1, 0, 0) \rightarrow (2, 0, 0) \rightarrow (3, 0, 0)$" indicates that: component A_1 degrades from state D_0 to state D_1 for the first time; the second degradation process occurs to component A_1 from state D_1 to state D_2; the third degradation is from state D_2 to state D_3 of component A_1; the available capacity of the system just satisfies system requirement according to the capacities of each degradation state. Therefore, it is not necessary to construct the MSDD for the fourth possible degradation process. For this path, components A_1 and A_2 are in operating state and component A_3 is in a warm standby state initially depending on system demand. When the first degradation process occurs, component A_3 switches to the operating state from the warm standby state to meet the system demand.

The probability of the system being in a particular state can be calculated by summing the probabilities of all disjoint paths which represent all possible combinations of degradation processes leading to the entire system in the particular state [13]. Thus, the system probability can be obtained by summing the occurrence probability of each path in the constructed MSDD. Therefore, it is essential to calculate the probability of each path in the MSDD. The occurrence probability of each path in the system MSDD can be obtained from Boolean function manipulation based on Shannon decomposition for decision diagrams [42]. More details for manipulation rules of Shannon decomposition for decision diagrams can be found in Reference [41].

Consider the leftmost path mentioned above for instance. Taking the start failure probability into consideration, the occurrence probability of this path is

$$Path_1(t) = (1 - p_3) \int_{\tau_0}^{t} \left\{ \int_{\tau_1}^{t} \left[\int_{\tau_2}^{t} f_2^{A_1}(\tau_3 - \tau_2) d\tau_3 \right] f_1^{A_1}(\tau_2 - \tau_1) d\tau_2 \right\} f_0^{A_1}(\tau_1) R_0^{A_2}(t) R_0^{A_3^w}(\tau_1) R_0^{A_3^o}(t - \tau_1) d\tau_1 \quad (7.4)$$

where t is the mission time, $\tau_0 = 0$ is the beginning time, $f_*^{\#}(t)$ is the probability density function (PDF), $F_*^{\#}(t)$ is the cumulative distribution function (CDF) and $R_*^{\#}(t) = 1 - F_*^{\#}(t)$ is the reliability function. τ_1 is an integral variable and located in (τ_0, t) indicating the first degradation time of A_1. τ_2 is located in (τ_1, t) representing the second degradation time of A_1.τ_3 is located in (τ_2, t) indicating the third degradation time of A_1. p_3 is the start failure probability of A_3 when switching from a warm standby state to an operating state.

The system reliability can be obtained by calculating the sum of the occurrence probability of each path. $R(t)$ is the time-varying reliability values of the DB-WSS. b represents the number of the path in the system MSDD representation.

$$R(t) = \sum_b Path_b(t) \quad (7.5)$$

7.3.3 Complexity Analysis of the Proposed MSDD-based Method

The complexity of the proposed MSDD-based method depends on both MSDD construction and reliability evaluation based on MSDD procedures [43]. Taking the degradation process of each step into account, the complexity of MSDD construction mainly lies in the number of terminals in the MSDD. For the reliability assessment procedures, the complexity mainly comes from the integration for occurrence probabilities of paths in system MSDD. Therefore, the computational effort of the proposed technique is supposed to be analyzed considering both the terminals in the MSDD construction and integration for occurrence probabilities, respectively.

According to the MSDD construction algorithm, the number of terminals in system MSDD is less than:

$$1+\sum_{s=1}^{m} N^s = (N^{1+m}-1)/(N-1) \quad (N>1, C_s \geq d) \tag{7.6}$$

It is denoted in Eq. (7.6) that the capacity of the degradation state, system demand and number of components will determine the number of terminals in system MSDD.

Specifically, if the system MSDD construction stops due to no further degradation, this scenario could be the worst scenario of the system MSDD construction. The number of terminals in the worst scenario is less than:

$$1+\sum_{s\in S} N^s = (N^{1+\sum_{i=1}^{N} M_i}-1)/(N-1) \quad (N>1) \tag{7.7}$$

where $S=\left\{1,2,\cdots,\sum_{i=1}^{N} M_i\right\}$. It is indicated in (7.7) that the number of terminals for system MSDD in the worst case is associated with the number of system components and the summation of the degradation states. However, it should be noted that, in general, the number of terminals in the proposed algorithm is much smaller than Eq. (7.6) since system MSDD construction stops when the available capacity of the system is insufficient to cover the required demand. Moreover, the algorithm can be further improved to construct unreliable paths if the number of reliable paths of the MSDD is much more than that of unreliable paths. Generally, the proposed MSDD-based technique does not need to enumerate all the possible combinations which in turn reduces the complexity of the MSDD construction.

Taking the DB-WSS in Fig. 7.7 as an example, the MSDD construction is stopped after the fourth degradation, and there are only 48 terminals, which result in 48 reliable paths in the system MSDD. However, if we enumerate all combinations, there are $(1+N)^{\sum_i M_i} = (1+3)^{3+3+3} = 4^9$ combinations for the example which is much larger than the number of paths in the proposed method.

For the process of MSDD construction, some subtrees can be merged to improve the algorithm performance if there are a number of subtrees that can be merged. In the given example, it can be seen in Fig. 7.7d that the number of terminals of MSDD considering the merged subtrees is 24 which is much less than that of the system MSDD representation without considering the merged subtrees. Moreover, in order to achieve the combination of subtrees, the nodes with identical system states are supposed to be searched utilizing the breadth-first search algorithm [44]. The breadth-first search algorithm may also incur additional computational effort for searching identical nodes indicating the same system states. Since in the worst case, the breadth-first search has to consider all paths to all possible nodes, the computational complexity of breadth-first search algorithm in this case is $O(V)$, where V indicates the number of terminals in the system MSDD [44].

The computational effort of integration for occurrence probabilities of each path in system MSDD is mainly derived from the solution of possible multiple integration. The multiplicity of integration depends on the number of degradation processes of each path. With the increase of the number of degradation processes, the integration becomes more and more complicated. If the number of degradation processes for a path in system MSDD is s, the multiplicity of integration for this path is s as well. Assuming that the integration region in a one-dimension is divided into g intervals, the integration region is supposed to be divided into $g^{\sigma \cdot s}$ sub integration regions when the situation comes to multi-dimension, where σ indicates the degree factor [45]. Therefore, the integration region can be divided into r sub integration regions where $O(r) = g^{\sigma \cdot s}$. Heap structure is usually utilized to sort these sub integration regions in which the sub integration region with maximal deviation is on the top of the heap. If deviation of a sub integration region does not meet the predetermined precision accuracy, the sub integration region is supposed to be split into two regions. Since the upper bound of computational complexity for heap operation has been proved to be $O(\log r)$ [46], the computational complexity to solve the integration utilizing the heap operation is $O(r \log r)$ [46]. It should be noted that system operation time impacts the integration region. With longer system operation time, the integration region becomes larger which results in a longer computation time of the proposed method.

Notice that, according to Eq. (7.3), the maximal number of degradation processes of a path is related to the number of system's components, the number of components' states, the capacities of components' states and system demand. With relatively small number of system components and states, the number of degradation processes for the system MSDD is limited and the integration could be solved. Moreover, if the difference between the system demand and the available capacity is not too large, and the difference of capacities for the component' adjacent states is not too small, the number of degradation processes would be small and the integration could be not so complicated. The proposed MSDD-based method is applicable to systems with relatively small number of system's components and components' states [43], or the systems with relatively small difference between system demand and available capacity, whereas the difference between capacities for the component' adjacent states is not too small. Furthermore, it should be noted that the occurrence probabilities of some paths with larger number of degradation processes are very small (close to zero). Therefore, the occurrence probabilities of paths with larger number of degradation processes can be neglected.

7.4 Numerical Studies

To illustrate the application of the proposed approach, four numerical studies are considered in this section.

7.4.1 Example 1: Exponential Distribution

Considering a DB-WSS with two components (A_1 and A_3) mentioned in Sect. 7.2, the transition time distributions for degradation processes for the two components are shown in Table 7.2. The transition time distributions for the degradation processes for each component are exponential distributions. In this case, the start failure probability of the warm standby component is 0.

Two different scenarios are studied in this case. The difference between the two scenarios is the system demand. Scenario A has a larger system demand with 7, while the system demand in Scenario B is 5. For a demand-based system, the different system demands determine the number of components in the operating state and warm standby state. When system demand is 7, the two components are all in operating states. When the system requirement is 5, component A_1 is an operating component and component A_3 is a warm standby component. The number of Monte Carlo simulations is set to be 100,000. In this case, the computer programs for the proposed methods were developed in Wolfram Mathematica 9.0 and were implanted on a computer with a 3.40 GHz processor.

There are 6 terminals and 15 terminals in the system MSDD for Scenarios A and B in Example 1, respectively. Moreover, the MSDD construction is stopped after the second and third degradation, respectively. Figure 7.8a and b illustrate the time-varying reliability values of the system utilizing the proposed approach and Monte Carlo simulation, respectively, for Scenarios A and B. It is illustrated that the result of the proposed method is approximate to that of the Monte Carlo simulation technique. The numerical comparisons of system reliability and relative differences for different approaches are given in Tables 7.3 and 7.4, which respectively display results with different time in Scenarios A and B. The relative differences indicate the minor differences between the results of the two methods. Table 7.5 presents the computation time of the proposed method and Monte Carlo simulation, which indicates that the computing time of the proposed method is shorter than that of the Monte Carlo simulation. The computation time of Scenario B is longer than that of Scenario A due to higher number of paths in the system MSDD.

In addition, comparing the reliability curves with different system demands, it is obvious that the larger the system demand is, the lower the system reliability will be.

Table 7.2 Degradation distributions for the components

Component	PDF in warm standby state	PDF in operating state
A_1	–	$f_0^{A_1} = f_1^{A_1} = f_2^{A_1}$ $= \lambda_1 \exp(-\lambda_1 t)$
A_3	$f_0^{A_3^w} = f_1^{A_3^w} = f_2^{A_3^w}$ $= \lambda_3^w \exp(-\lambda_3^w t)$	$f_0^{A_3} = f_1^{A_3} = f_2^{A_3}$ $= \lambda_3^o \exp(-\lambda_3^o t)$

$\lambda_1 = 1/100, \lambda_3^w = 1/300, \lambda_3^o = 1/150$ (/day)

Fig. 7.8 The system reliability of the proposed method and Monte Carlo simulation technique for Example 1

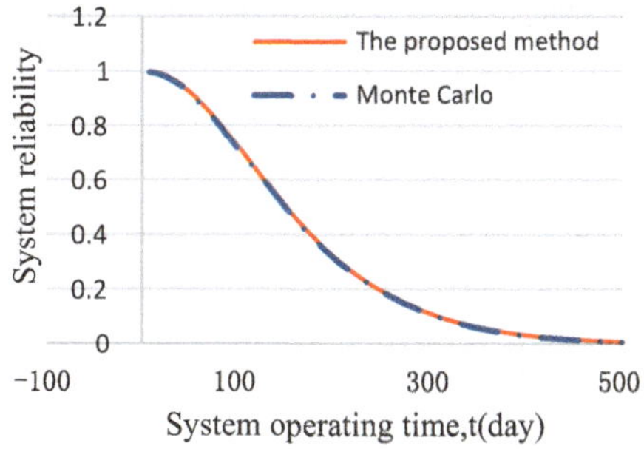

(a) The system reliability of the proposed method and Monte Carlo simulation technique in Scenario A (system demand is 7)

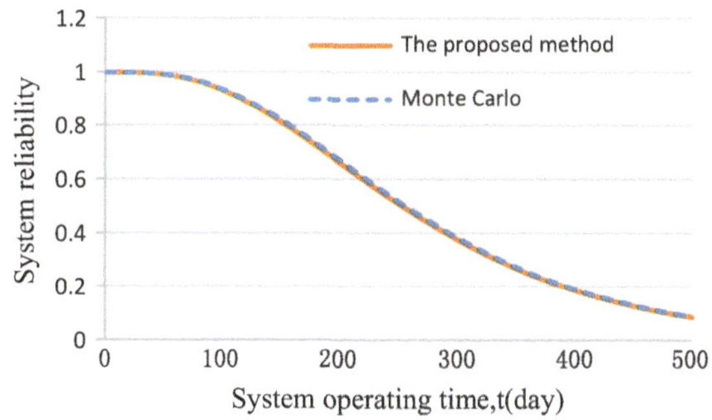

(b) The system reliability of the proposed method and Monte Carlo simulation technique in Scenario B (system demand is 5)

Table 7.3 Comparison of reliability for different methods in scenario A

Time (day)	System reliability for Scenario A in Example 1		Relative differences of different methods (%)
	Proposed method	Monte Carlo	
50	0.9235	0.9238	0.03
100	0.7240	0.7259	0.26
200	0.3211	0.3266	1.68
500	0.0093	0.0097	4.12

7.4.2 *Example 2: Weibull Distribution*

To illustrate the application of the proposed method on non-exponential distribution for degradation processes, a DB-WSS of three components (A_1, A_2 and A_3) with Weibull distribution is considered for its wide applications in reliability engineering.

Table 7.4 Comparison of reliability for different methods in scenario B

Time (day)	System reliability for Scenario B in Example 1		Relative differences of different methods (%)
	Proposed method	Monte Carlo	
50	0.9917	0.9931	0.14
100	0.9333	0.9385	0.55
200	0.6610	0.6730	1.78
500	0.0864	0.0884	2.26

Table 7.5 Comparison of computation time for different scenarios in Example 1

Computation time (s)	Scenario A (d = 7)	Scenario B (d = 5)
Proposed method	8.383	22.292
Monte Carlo	130.670	142.987

Table 7.6 Weibull distribution parameters for the degradation process

Component	For warm standby state		For operating state	
	Scale parameter	Shape parameter	Scale parameter	Shape parameter
A_1	–		200	2
A_2	–		150	2
A_3	300	1.5	150	1.5

Table 7.6 lists scale parameter μ and shape parameter η for the Weibull distribution of the degradation time. It is assumed that the degradation process follows the same distribution for the same components. The CDF for the Weibull distribution is

$$F = 1 - \exp\left[-(t/\mu)^{\eta}\right] \tag{7.8}$$

The system demand is 10. To make a comparison, different start failure probabilities of the warm standby component A_3 are taken into account.

In this case, the computer programs were implemented on the same platform as used in Example 1. There are 48 terminals resulting in 48 reliable paths for the MSDD construction. The MSDD construction is stopped after the fourth degradation which leads to a longer computation time. The computation time in this example whose operation time is 200 days is 36.237 s. As shown in Fig. 7.9 and Table 7.7, the system reliability increases with a decrease in the start failure probability values.

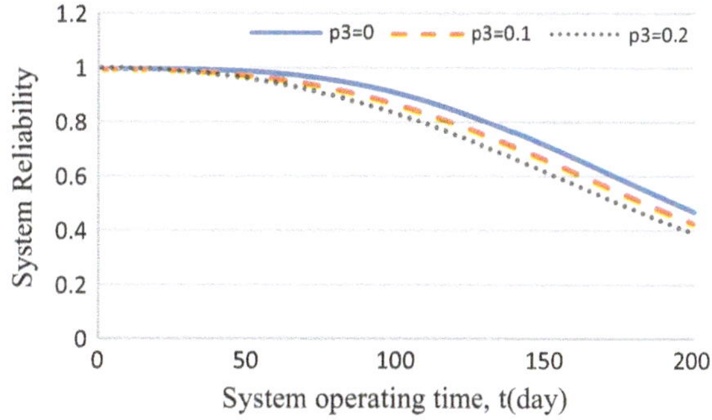

Fig. 7.9 System reliability of the proposed method with different values of start failure probability

Table 7.7 System reliability with different start failures

t (day)	p3 = 0	P3 = 0.1	P3 = 0.2
20	0.999719	0.997007	0.994296
40	0.996379	0.986225	0.976071
100	0.910803	0.869658	0.828513
200	0.471692	0.43074	0.389789

7.4.3 *Example 3: DB-WSS with 10 Components*

To illustrate the application of the proposed approach on large-scale systems, a DB-WSS with 10 components is considered. The capacities of each state for components and the distributions of the degradation transition are given in Table 7.8.

There are 8 online components and 2 warm standby components in the DB-WSS due to the state capacities of each component and the system demand requirement. The MSDD representation is not given here because of the paragraph space limitation. There are 165 terminals resulting in 165 reliable paths for the MSDD construction. The MSDD construction is stopped after the third degradation. The computation time of the proposed method in this case with 100 days is 6.187 s. The system reliability subjected to different start failure probabilities is illustrated in Fig. 7.10. It is clearly seen that the system reliability is influenced by the start failures.

Table 7.8 System configurations for a DB-WSS with 10 components

Component/State	D_0	D_1	D_2	PDF for warm standby state	PDF for operating state
$A_1, A_2, \cdots, A_8$	10	5	0	–	$f_0 = f_1 = \lambda_1 \exp(-\lambda_1 t)$
A_9, A_{10}	5	2	0	$f_0^w = f_1^w = \lambda_2^w \exp(-\lambda_2^w t)$	$f_0 = f_1 = \lambda_2 \exp(-\lambda_2^o t)$
$\lambda_1 = 1/100, \lambda_2^w = 1/300, \lambda_2^o = 1/150$,system demand = 80					

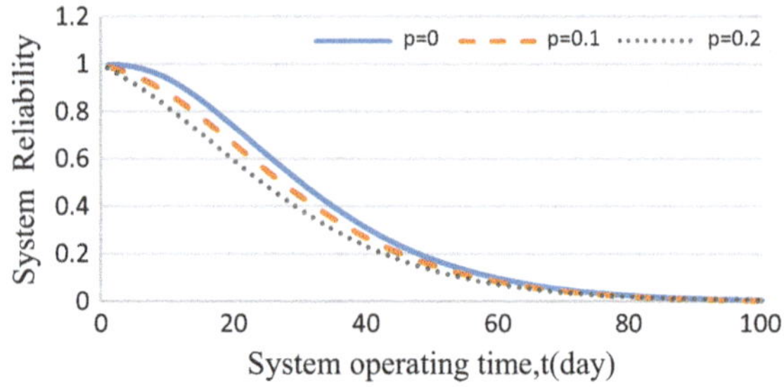

Fig. 7.10 System reliability of the DB-WSS (10 components) with different start failure probabilities

To make a comparison between exponential distribution and non-exponential distribution, a system with Weibull distribution is considered in this section. The parameters of the Weibull distribution for components are described in Table 7.9.

The system MSDD construction procedure of the components with Weibull distribution is identical to that of the components with exponential distribution. Obviously, the differences lie in the reliability result and computation time, which result from the integral formula for each path of the system MSDD. Figure 7.11 demonstrates the system reliability of DB-WSS with 10 components whose degradation time transitions follow Weibull distributions. The comparison of computation time for 10 components DB-WSS with exponential distributions and Weibull distributions is provided in Table 7.10. It indicates that the computation time of the system with

Table 7.9 Weibull distribution parameters for the degradation process

Component	For warm standby state		For operating state	
	Scale parameter	Shape parameter	Scale parameter	Shape parameter
$A_1, A_2, \cdots, A_8$	–		200	2
A_9, A_{10}	300	1.5	150	1.5

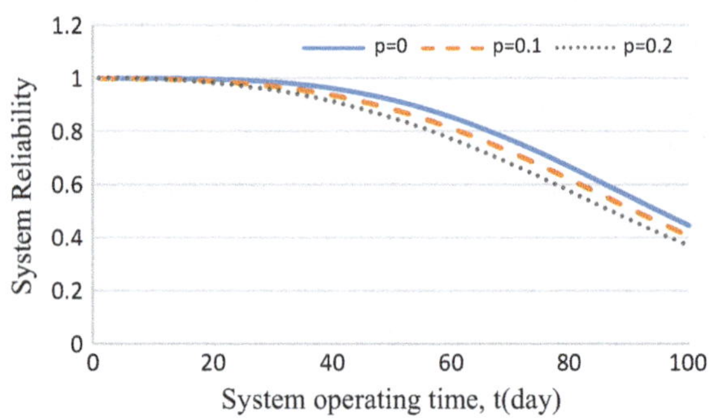

Fig. 7.11 Reliability of DB-WSS (10 components) with Weibull distributions

Table 7.10 Comparison of computation time of the 10-component system with different distributions

Computation time (s)	Exponential distribution	Weibull distribution
p = 0	5.521	9.991
p = 0.1	5.524	10.827
p = 0.2	5.996	11.706

Table 7.11 System configurations of the power generation system

Component/State	Generation of different states (MW)			Transition rates (/day)	
				Warm standby state	Operating state
A_1, A_2, A_3	20	10	0	–	1/146
A_4, A_5	10	5	0	1/365	2/365

Weibull distributions is much longer than that of the system with exponential distributions due to the complicated integral for Weibull distributions. Moreover, the failure rates of warm standby components have little influence on the computation time.

7.4.4 Example 4: DB-WSS for the Power Generation System

To illustrate the practical significance of the proposed method, a case study has been provided based on DB-WSS for the power generation system.

The generating units in IEEE RBTS [47] are modified to validate the proposed method. Three 20-MW thermal generating units and two 10-MW thermal generating units are installed in the WSS with the load of 64 MW. It is assumed that there are three states for generating units, and the transition rates of the same generating unit are equivalent. Parameters including generation of each state and the corresponding transition rates of generating units are displayed in Table 7.11. The start failures of the warm standby generating unit are assumed as 0, 0.1 and 0.2, respectively.

Obviously, for the DB-WSS of the power system, three 20-MW generating units and one 10-MW generating unit are in operating states, while the other 10-MW generating unit is in a warm standby state initially. There are 28 terminals resulting in 28 reliable paths for the MSDD construction. The MSDD construction is also stopped after the third degradation. The computation times of the proposed MSDD-based method for 100 days of different start failures are 4.499 s, 4.465 s and 4.652 s, respectively, which resulted from fewer reliable paths of MSDD construction. The reliability of the DB-WSS for power generation system is depicted in Fig. 7.12, which shows that system reliability decreases with an increase in the system operating time.

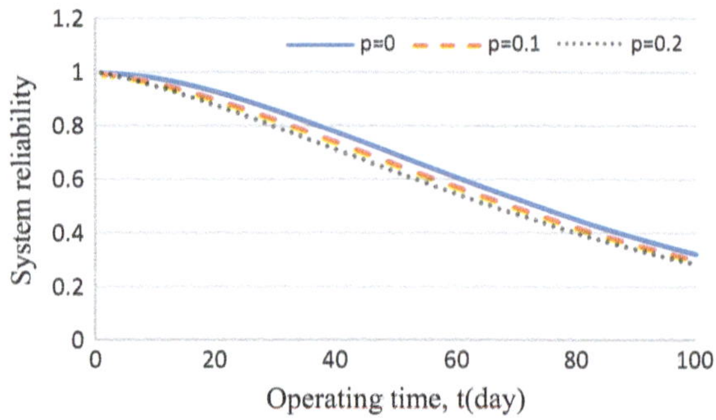

Fig. 7.12 System reliability of the DB-WSS for the power generation system

7.5 Conclusion

This chapter proposed the MSDD-based technique to evaluate the reliability for DB-WSS with degradation process considering start failure probabilities. For DB-WSS, the warm standby components have different transition time distributions for degradation processes before and after they are converted to operating states from warm standby states. The degradation process is illustrated as a multi-state model. The existing methods have various limitations such as being applied to binary-state systems or assuming exponential state transition distributions for all system components. The proposed approach can handle arbitrary types of transition time distributions for degradation processes for DB-WSS. From the computational complexity analysis, the proposed method is applicable to systems with relatively small number of system's components and components' states, or the systems with relatively small difference between system demand and available capacity, whereas the difference between capacities for the component' adjacent states is not too small. Several numerical studies are given to illustrate the application of the proposed approach. It is shown that the MSDD-based technique is an effective and accurate approach to handle DB-WSS with degradation process.

The further researches can be devoted to the approximation of system with similar states utilizing fuzzy theory to improve the algorithm performance when the fuzzy clustering does not reduce the accuracy of the results considerably [48]. Moreover, simulation methods can be developed to analyze the reliability for the system with large number of degradation processes where the multiple integration could not be directly calculated by numerical integration.

References

1. G. Levitin, L. Xing, Y. Dai, "Reliability of non-coherent warm standby systems with reworking," IEEE Transactions on Reliability, vol. 64, no. 1, pp. 444–453, March 2015.
2. W. Kuo, M. J. Zuo, "Optimal Reliability Modeling, Principles and Applications," John Wiley & Sons, Inc, first edition, 2002.

3. S.V. Amari, H. Pham, R. B. Misra, "Reliability Characteristics of k-out-of-n Warm Standby Systems," IEEE Transactions on Reliability, vol. 61, no. 4, pp. 1007–1018, Dec. 2012.
4. M. Sun, F. Chen, "Research on availability of virtual machine hot standby based on Xen," in Software Intelligence Technologies and Applications & International Conference on Frontiers of Internet of Things 2014, 2014, pp. 330–335.
5. L. Yuan, X. Y. Meng, "Reliability analysis of a warm standby repairable system with priority in use," Applied Mathematical Modelling, vol. 35, no. 9, pp. 4295–4303, 2011.
6. G. Levitin, L. Xing, H. Ben-Haim, Y. Dai, "Effect of Failure Propagation on Cold vs. Hot Standby Tradeoff in Heterogeneous 1-Out-of-n: G Systems," IEEE Transactions on Reliability, vol. 64, no. 1, pp. 410–419, March 2015.
7. S. V. Amari, G. Dill, "Redundancy optimization problem with warm-standby redundancy," in 2010 Proceedings-Annual IEEE Reliability and Maintainability Symposium (RAMS), 2010, pp. 1–6.
8. W. Y. Yun, J. H. Cha, "Optimal design of a general warm standby system," Reliability Engineering & System Safety, vol. 95, no. 8, pp. 880–886, 2010.
9. R. Billinton, R. N Allan, "Reliability evaluation of power systems," Springer Science & Business Media, 2013.
10. A. Filieri, C. Ghezzi, V. Grassi, R. Mirandola. "Reliability analysis of component-based systems with multiple failure modes," Component-Based Software Engineering, Springer Berlin Heidelberg, pp. 1–20, 2010.
11. M. L. J. Rhodin, "Reliability Calculations for Complex Systems," Department of Electrical Engineering, Linkö¨pings University, Sweden, pp. 75–85, 2011.
12. R. Peng, Q. Zhai, L. Xing, J. Yang, "Reliability of demand-based phased-mission systems subject to fault level coverage," Reliability Engineering & System Safety, vol. 121, no. 1, pp. 18–25, 2014.
13. Q. Zhai, R. Peng, L. Xing, J. Yang, "Reliability of demand-based warm standby systems subject to fault level coverage," Applied Stochastic Models in Business and Industry, vol. 31, no. 3, pp. 380–393, 2015.
14. T. Jin, Y. Yu, E. Elsayed, "Reliability and quality control for distributed wind/solar energy integration: a multi-criteria approach," IIE Transactions, vol. 47, no. 10, pp. 1122–1138, 2015.
15. Z. Wang, H. Z. Huang, Y. Li, N. C. Xiao, "An approach to reliability assessment under degradation and shock process," IEEE Transactions on Reliability, vol. 60, no. 4, pp. 852–863, Dec. 2011.
16. A. Lianianski, G.Levitin, "Multi-State System Reliability: Assessment, Optimization and Application," New York: World Scientific Publishing Co. Pte. Ltd, 2003.
17. Y. F. Li, E. Zio, Y. H. Lin, "A multistate physics model of component degradation based on stochastic Petri nets and simulation," IEEE Transactions on Reliability, vol. 61, no. 4, pp. 921–931, Dec. 2012.
18. G. K. Chana, S. Asgarpoor, "Optimum maintenance policy with Markov processes," Electric Power Systems Research, vol. 76, pp. 452–456, 2006.
19. S. H. Sim J. Endrenyi, "A failure-repair model with minimal and major maintenance," IEEE Transactions on Reliability, vol. 42, no. 1, pp. 134–139, March 1993.
20. M. Black, A. T. Brint, J. R. Brailsford, "A semi-Markov approach for modelling asset deterioration," Journal of the Operational Research Society, vol. 56, pp. 1241–1249, 2005.
21. J. Kim, V. Makis, "Optimal maintenance policy for a multi-state deteriorating system with two types of failures under general repair," Computers and Industrial Engineering, vol. 57, pp. 298–303, 2009.
22. J. P. Kharoufeh, S. M. Cox, "Stochastic models for degradation-based reliability," IIE Transactions, vol. 37, no. 6, pp. 533–542, 2005.
23. G. Fishman, "Monte Carlo: concepts, algorithms, and applications," Springer Science & Business Media, 2013.
24. W. Li, "Reliability assessment of electric power systems using Monte Carlo methods," Springer Science & Business Media, 2013.

25. E. Zio, "The Monte Carlo simulation method for system reliability and risk analysis," London: Springer, 2013.
26. Y. Ding, P. Wang, L. Goel, P. C. Loh, Q. Wu, "Long-term reserve expansion of power systems with high wind power penetration using universal generating function methods," IEEE Transactions on Power System, vol. 26, no. 2, pp. 766–774, May 2011.
27. A. Shrestha, L. Xing, Y. Dai, "Decision diagram based methods and complexity analysis for multi-state systems," IEEE Transactions on Reliability, vol. 59, no. 1, pp. 145–161, March 2010.
28. Q. Zhai, R. Peng, L. Xing, Y. Jun, "Binary decision diagram-based reliability evaluation of k-out-of- (n + k) warm standby systems subject to fault-level coverage," Proceedings of the Institution of Mechanical Engineers, Part O: Journal of Risk and Reliability, vol. 227, no.5, pp. 540–548, 2013.
29. Y. Mo, L. Xing, S.V. Amari, "A Multiple-Valued Decision Diagram Based Method for Efficient Reliability Analysis of Non-Repairable Phased-Mission Systems," IEEE Transactions on Reliability, vol. 63, no.1, pp. 320–330, March 2014.
30. Y. Mo, "New Insights Into the BDD-Based Reliability Analysis of Phased-Mission Systems," IEEE Transactions on Reliability, vol. 58, no. 4, pp. 667–678, Dec. 2009.
31. Ding, L. Cheng, Y. Zhang, Y. Xue, "Operational reliability evaluation of restructured power systems with wind power penetration utilizing reliability network equivalent and time-sequential simulation approaches," Journal of Modern Power Systems and Clean Energy, vol. 2, no. 4, pp. 329–340, 2014.
32. G. Levitin, "Universal Generating Function and Its Applications", New York: Springer, 2005.
33. Y. Ding, C. Singh, L. Goel, J. Østergaard, P. Wang, "Short-term and medium-term reliability evaluation for power systems with high penetration of wind power," IEEE Transactions on Sustainable Energy, vol. 5, no. 3, pp. 896–906, July 2014.
34. G. Levitin, "The universal generating function in reliability analysis and optimization," London: Springer, 2005.
35. Y. Ding, A. Lisnianski, "Fuzzy universal generating functions for multi-state system reliability assessment," Fuzzy Sets and Systems, vol. 159, no. 3, pp. 307–324, 2008.
36. F. Li, E. Zio, "A multi-state model for the reliability assessment of a distributed generation system via universal generating function," Reliability Engineering & System Safety, vol. 106, no. 5, pp. 28–36, 2012.
37. S.Daichman, A. Lisnianski, "On aging components impact on multi-state water cooling system: Lz-transform application for availability assessment," in 2013 IEEE International Conference on Digital Technologies (DT), 2013, pp. 156–161.
38. I. Frenkel, A. Lisnianski, "Assessing Water Cooling System Performance: Lz-Transform Method," in IEEE 2013 Eighth International Conference on Availability, Reliability and Security (ARES), 2013, pp. 737–742.
39. A. Lisnianski, Y. Ding, "Redundancy analysis for repairable multi-state system by using combined stochastic processes methods and universal generating function technique," Reliability Engineering & System Safety, vol. 94, no. 11, pp. 1788–95, 2009.
40. I. Frenkel, A. Lisnianski, "Performance Determination for MSS Manufacturing System by Lz-Transform and Stochastic Processes Approach," in IEEE 2014 Ninth International Conference on Availability, Reliability and Security (ARES), 2014, pp. 387–392.
41. O. Tannous, L. Xing, J. B. Dugan, "Reliability analysis of warm standby systems using sequential BDD," in 2011 Proceedings-Annual Reliability and Maintainability Symposium (RAMS), 2011, pp. 1–7.
42. R. E. Bryant. "Graph-based algorithms for Boolean function manipulation," IEEE Transactions on Computers, vol. 35, no. 8, pp. 677–691, Aug. 1986.
43. Q. Zhai, L. Xing, R. Peng, J. Yang, "Multi-Valued Decision Diagram-Based Reliability Analysis of k-out-of-n Cold Standby Systems Subject to Scheduled Backups," IEEE Transactions on Reliability, vol. 64, no. 4, pp. 1310–1324, Dec. 2015.
44. B. Awerbuch and R. Gallager, "A new distributed algorithm to find breadth first search trees," IEEE Transactions on Information Theory, vol. 33, no. 3, pp. 315–322, May 1987.

45. A. Quarteroni, R. Sacco, F. Saleri, in Numerical mathematics, second edition, Springer, 2006.
46. M. L. Fredman, H. S. Thomas, "Refined complexity analysis for heap operations," Journal of Computer and System Sciences, vol. 35, pp. 269–284, 1987.
47. R. Billinton, S. Kumar, N. Chowdhury, et al., "A reliability test system for educational purposes-basic data," IEEE Transactions on Power Systems, vol. 4, no. 3, pp. 1238–1244, Aug. 1989.
48. Y. Ding, M. J. Zuo, A. Lisnianski and W. Li, "A Framework for Reliability Approximation of Multi-State Weighted k-out-of-n Systems," IEEE Transactions on Reliability, vol. 59, no. 2, pp. 297–308, June 2010.

Chapter 8
Reliability of Demand-Based Warm Standby System with Common Bus Performance Sharing

Redundancy techniques have been extensively utilized to enhance the reliability of engineering systems. There are three different types of standby techniques, cold, hot, and warm. Warm standby is adopted for less energy consumption and shorter leading time compared with hot standby and cold standby, respectively. Besides redundancy, performance sharing is another strategy to enhance system reliability, where the subsystems with sufficient performance can share the surplus performance with other subsystems with deficient performance. We consider a demand-based warm standby system with a common bus performance sharing mechanism, where the system subsystems can share performance through the common bus and each subsystem can be configured with warm standby components in order to meet its demand. To be more general, the imperfect switching for the activations of warm standby components is also considered. Moreover, the multi-valued decision diagram technique is developed to analyze the reliability for the proposed model. The proposed technique can handle systems whose time-to-failure distributions can follow arbitrary distributions in addition to the common utilized exponential distributions. Numerical studies are provided to validate the proposed model and technique.

8.1 Introduction

Redundancy techniques have been extensively applied to engineering systems to enhance system reliability. There are three types of redundancies: hot standby, cold standby, and warm standby [1]. Hot standby implies a system consisting of online components while other components function synchronously as backup [2]. The hot standby components can be put into operation immediately when system emergency occurs with more energy consumption compared with cold and warm standby. In cold standby systems, the backup components are activated upon a system emergency and it may take long leading time to get into work [3]. The warm standby is a balance between hot standby and cold standby. In warm standby system, less energy

R. Peng et al., *Reliability Modelling and Optimization of Warm Standby Systems*,
https://doi.org/10.1007/978-981-16-1792-8_8

consumption compared with hot standby and shorter leading time compared with cold standby can be achieved [4]. The failure rates of hot standby components are the same as those of online components while the failure rates of cold standby components are much lower than those of online components. The failure rates of warm standby components are lower than the online components since they suffer from less system pressure [5].

The demand-based warm standby system (DBWSS) denotes a system where the modes of its components are determined by the system demand [6–8]. If the demand is satisfied with certain components, these components are in online modes, while the remaining components are in warm standby modes.

Moreover, in many practical engineering systems, for example, power systems and communication systems, each subsystem of the system has to satisfy its individual demand [9]. Moreover, if its individual demand can be satisfied, the surplus performance of the subsystem can be shared with other subsystems which are suffering from performance deficiency. This kind of system is performance sharing system which is first proposed in [10] where the transmission of surplus performance is limited from the standby unit to the online unit. It was extended in [11] as common bus performance sharing system by modifying the system with arbitrary number of components and the surplus performance can be shared multi-directionally.

It is indicated that the surplus performance can be redistributed between the components and the constraint of the redistribution is the capacity of the transmission system. Furthermore, the system is improved by complicated system structure with series–parallel system [12] instead of simple series system [10]. Time-varying reliability evaluation of systems with common bus performance sharing is proposed in [13] instead of stead-state reliability. However, it should be noted that redundancy technique can be applied to each subsystem with parallel structure. In [14], the systems are extended to systems with two performance sharing groups. The effect of loading and protection of external factor for systems with performance sharing is investigated in [15]. The reliability of non-repairable phased-mission systems considering common bus performance sharing is analyzed in [16]. Motivated by practical engineering systems, the DBWSS with common bus performance sharing is proposed. Furthermore, it should be noted that for the warm standby components, they may suffer from start failures or imperfect switches when activated. The constant start failure probability is generally adopted in reliability engineering for simplicity.

Considering the reliability analysis techniques for common bus performance sharing systems, the universal generating function methods are adopted in [11, 12]. The combined stochastic process and universal generating function method is utilized to evaluate the time-varying reliability of the performance sharing system [13]. However, these methods do not consider different failure rates for warm standby components and online components and might not be directly applied to warm standby systems. Moreover, the state probabilities are supposed to be known for the reliability evaluation when utilizing universal generating function methods or the time-to-failure distributions are assumed to be exponential when adopting Markov process model [17, 18]. Recent years, decision diagram approaches have been extensively utilized in the reliability analysis of engineering systems, for example, redundancy systems [19, 20] phased mission systems [21–23]. The multi-valued decision diagram (MDD) techniques in [24] have been extended from binary decision diagram methods [25] for the reliability analysis of complex engineering systems. The main advantage of the decision diagram lies in handling systems with arbitrary time-to-failure distributions. This method is significantly extended to analyze the reliability of DBWSS with common bus performance sharing.

A non-repairable DBWSS with common bus performance sharing is proposed where the surplus performance of the subsystems can be redistributed to other subsystems with performance deficiency. Moreover, the redistribution capacity is limited by the capacity of the common bus. The start failures of warm standby components are incorporated in the proposed model and technique. The MDD based technique is developed to deal with the proposed model where the systems can follow arbitrary time-to-failure distributions in addition to the commonly adopted exponential distribution. The time-varying reliabilities of the proposed systems are presented.

8.2 Model Description for the DBWSS with Common Bus Performance Sharing

As shown in Fig. 8.1, the model for the DBWSS with common bus performance sharing is proposed. The DBWSS is composed of N demand-based warm standby subsystems where subsystem j must satisfy the corresponding demand L_j. The set of components in subsystem j is G_j. The system consists of M binary-state components with perfect functioning and complete down. The performance of the binary component g_i is P_i which equals to the capacity *HiHi or 0*. If the component suffers a failure with predesigned probability density function (PDF), the component is completely down. The numbers of online components and warm standby components of the subsystem j are determined by its individual demand L_j. If the total capacity of former components in G_j satisfies the predetermined demand L_j, the remaining components of the subsystem are in the warm standby mode.A_i^o indicates that component A_i is in online mode and A_i^w denotes that component A_i is in the warm standby mode. When a failure occurs to arbitrary component, one or more components in the warm standby mode will be activated to online mode to meet the system demand and the performance deficiency resulted from the component failure is made up by the activated warm standby components. In the proposed model, the failed online component is assumed to be not repaired in the system operation and it is removed from the system [26].

The warm standby component is activated in a predesigned sequence as their indexes[6]. Moreover, if a performance deficiency occurs in a subsystem, the warm standby components belonging to this subsystem are prioritized to be activated. For example, if an online component fails and the resulted performance deficiency is 5, and there are three standby components, A_1, A_2 and A_3 with performance 4, 4 and 10, respectively. In this scenario, standby components A_1 and A_2 will be activated under the assumption that warm standby components are activated in a predesigned sequence as their indexes. Supposing the transmission capacities are always *Inf*, if later on another online system fails and the performance deficiency is 5, standby component A_3 will be activated because the previously activated components A_1 and A_2 cannot meet the system demand.

The demand-based warm standby subsystems are connected with a common bus performance sharing system, where the transmission capacity is predesigned as C. If the performance of subsystem j exceeds the demand L_j, the surplus performance can be transmitted to any other subsystems which are suffering from performance deficiency. However, the transmission is limited by the transmission capacity of the common bus performance sharing system. The DBWSS suffers a system failure if and only if at least one of the subsystems cannot satisfy the demand.

The total surplus performance of the DBWSS can be formulated as

$$S = \sum_{j=1}^{N} S_j = \sum_{j=1}^{N} \max\left(\sum_{g_i \in G_j} P_i - L_j, 0\right). \tag{8.1}$$

The total performance deficiency of the DBWSS can be expressed as

$$D = \sum_{j=1}^{N} D_j = \sum_{j=1}^{N} \max\left(L_j - \sum_{g_i \in G_j} P_i, 0\right). \tag{8.2}$$

Considering the common bus performance sharing system, the subsystems with performance deficiency can be compensated by the subsystems with surplus performance where no more than S amount of performance can be provided. Therefore, the amount of performance which is supposed to be transmitted is min{S, D}. Moreover, owing to the limited transmission capacity of the common bus performance sharing system, the amount of performance which can be transmitted in the entire system can be formulated as Eq. (8.3).

$$\begin{aligned} Z &= \min(S, D, C) \\ &= \min\left(\begin{array}{c} \sum_{j=1}^{N} \max\left(\sum_{g_i \in G_j} P_i - L_j, 0\right), \\ \sum_{j=1}^{N} \max\left(L_j - \sum_{g_i \in G_j} P_i, 0\right), C \end{array}\right). \end{aligned} \tag{8.3}$$

It should be noted that S and D are statistically dependent, where S and C as well as D and C are statistically dependent.

The performance deficiency of the total system after the redistribution is

$$\begin{aligned} D' &= D - Z = D - \min(S, D, C) \\ &= \max(0, D - \min(S, C)) \\ &= \max\left(\begin{array}{l} 0, \sum_{j=1}^{N} \max\left(L_j - \sum_{g_i \in G_j} P_i, 0 \right) - \\ \min\left(\sum_{j=1}^{N} \max\left(\sum_{g_i \in G_j} P_i - L_j, 0 \right), C \right) \end{array} \right). \end{aligned} \quad (8.4)$$

The remaining unused surplus performance can be expressed as Eq. (8.5).

$$\begin{aligned} S' &= S - Z = S - \min(S, D, C) \\ &= \max(0, S - \min(D, C)) \\ &= \max\left(\begin{array}{l} 0, \sum_{j=1}^{N} \max\left(\sum_{g_i \in G_j} P_i - L_j, 0 \right) - \\ \min\left(\sum_{j=1}^{N} \max\left(L_j - \sum_{g_i \in G_j} P_i, 0 \right), C \right) \end{array} \right). \end{aligned} \quad (8.5)$$

The system is considered to be reliable if there is no performance deficiency for each subsystem after the redistribution. The reliability of the DBWSS with performance sharing can be presented as Eq. (8.6).

$$\begin{aligned} R &= \Pr\{D' = 0\} \\ &= \Pr\left\{ \max\left(\begin{array}{l} 0, \sum_{j=1}^{N} \max\left(L_j - \sum_{g_i \in G_j} P_i, 0 \right) - \\ \min\left(\sum_{j=1}^{N} \max\left(\sum_{g_i \in G_j} P_i - L_j, 0 \right), C \right) \end{array} \right) = 0 \right\}. \end{aligned} \quad (8.6)$$

Specifically, if there is no common bus performance sharing system, i.e., C = 0, the proposed DBWSS becomes a series–parallel system. If the transmission capacity is not constrained, i.e., $C = \infty$, the proposed systems is a parallel system.

When $C = 0$, the updated performance deficiency after redistribution can be simplified as Eq. (8.7).

$$\begin{aligned} D' &= D - \min(S, D, C) = D - \min(S, D, 0) \\ &= D = \sum_{j=1}^{N} \max\left(L_j - \sum_{g_i \in G_j} P_i, 0\right) \end{aligned} \tag{8.7}$$

The system reliability is formulated as Eq. (8.8).

$$\begin{aligned} R &= \Pr\{D' = 0\} = \Pr\left\{\sum_{j=1}^{N} \max\left(L_j - \sum_{g_i \in G_j} P_i, 0\right) = 0\right\} \\ &= \Pr\left\{\prod_{j=1}^{N}\left(L_j \le \sum_{g_i \in G_j} P_i\right)\right\} \end{aligned} \tag{8.8}$$

When $C = \infty$, the updated performance deficiency after redistribution can be reduced to Eq. (8.9).

$$\begin{aligned} D' &= D - \min(S, D, \infty) = \max(0, D - S) \\ &= \max\left(\begin{array}{l} 0, \sum_{j=1}^{N} \max\left(L_j - \sum_{g_i \in G_j} P_i, 0\right) - \\ \sum_{j=1}^{N} \max\left(\sum_{g_i \in G_j} P_i - L_j, 0\right) \end{array}\right) \\ &= \max\left(0, \sum_{j=1}^{N} L_j - \sum_{j=1}^{N} \sum_{g_i \in G_j} P_i\right) \\ &= \max\left(0, \sum_{j=1}^{N} L_j - \sum_{i=1}^{M} P_i\right) \end{aligned} \tag{8.9}$$

Then, the system reliability is presented as Eq. (8.10).

$$\begin{aligned} R &= \Pr\{D' = 0\} \\ &= \Pr\left\{\max\left(0, \sum_{j=1}^{N} L_j - \sum_{i=1}^{M} P_i\right) = 0\right\} \\ &= \Pr\left\{\sum_{j=1}^{N} L_j \le \sum_{i=1}^{M} P_i\right\} \end{aligned} \tag{8.10}$$

Note that when a component fails and warm standby components are activated, the total surplus performance of the system changed. It would not change the failure time of other system components except the just activated warm standby components. As there is a chance that the next failure comes from these just activated components, the system reliability regarding the following failures is changed.

An illustrative example is provided to make the proposed model a better understanding. Considering a power generation system with two subsystems, each of which consists of two generating units, as presented in Fig. 8.2. The generation capacity of each components, the demand of each subsystem as well as the transmission capacity are presented in Table 8.1.

In initial, all components are in the up states. According to the capacities and demand, in subsystem 1, component A_1 is in online mode while component A_2 is in warm standby mode; for subsystem 2, both componentsA_3 and A_4 are in online modes. If A_1 suffers from component failure, component A_2 will be activated to satisfy the demand of subsystem 1. If a failure occurs to component A_3, component A_2 will be

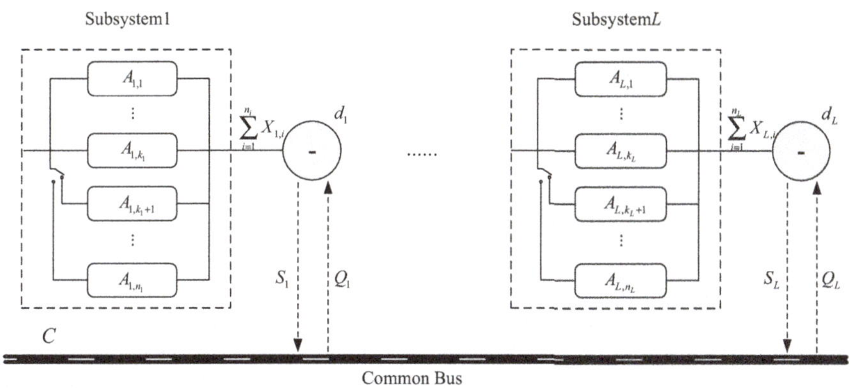

Fig. 8.1 The DBWSS with common bus performance sharing

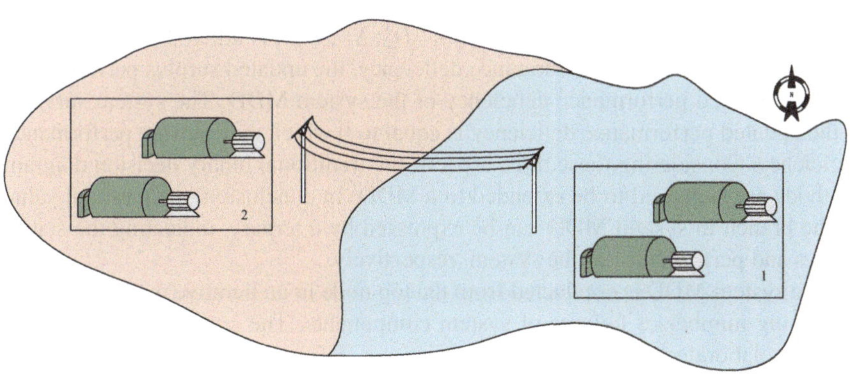

Fig. 8.2 The structure of the power generating system

Table 8.1 The information about the power generating system

Component	A_1	A_2	A_3	A_4	L_1	L_2	C
Capacity/demand	10	10	5	5	10	8	10

activated from warm standby mode to online mode and the surplus performance will be transferred from subsystem 1 to subsystem 2 through the common bus. Under this circumstance, the surplus performance of subsystem 1, the performance deficiency of subsystem 2, and the transmission capacity constraint are 10, 3, and 10, respectively. Therefore, the actual amount of surplus performance transmitted to subsystem 2 is equal to 3. Furthermore, if another failure occurs to component A_1 or A_2, the power generation system will fail for providing inadequate generation.

8.3 Time-Varying Reliability Evaluation Based on MDD

The decision diagram techniques have been successfully utilized in the reliability evaluation of warm standby systems [19]. The MDD technique is developed to be applied to the DBWSS with common bus performance sharing.

8.3.1 The Construction of System MDD

According to the system description in the model description, the following aspects should be considered when constructing MDD for DBWSS with common bus performance sharing. First, the vector $\mathbf{B}_x^y$ is utilized for indicating the states of components ("1" represents perfect functioning and "0" represents complete down) of the *xth* branch for the *yth* possible failure. Moreover, the vector $\mathbf{E}_x^y$ represents whether the components are in online modes or warm standby modes ("1" indicates online mode while "0" indicates warm standby mode) at the xth branch for the yth possible failure. Furthermore, the vector $\mathbf{Q}_x^y = \left[S_x^y, D_x^y, S_x^{\prime y}, D_x^{\prime y}\right]$ is utilized to represent the surplus performance, the performance deficiency, the updated surplus performance, and the updated performance deficiency of the system MDD. The system survives if the updated performance deficiency is equal to 0 which indicates no performance deficiency. Considering these three aspects, the traditional binary decision diagram methods are supposed to be extended to a MDD. In conclusion, the terminal value of the branch in system MDD can be expressed by a ternary, indicating the states, modes and performance of the system, respectively.

The system MDD is conducted from the top node in an iterative method with the increasing number of failures of system components. The construction of system MDD is elaborated in the following steps.

Step 1: Construct the top node of the system MDD.

At first, the top node of the MDD indicates that all components are in perfect functioning states and the modes of the components are determined by their performances and the demand of each subsystem. The values of the top node can be formulated as Eq. (8.11).

$$\begin{cases} \mathbf{B}_1^0 = \underbrace{(1,\cdots,1)}_{M} \\ \mathbf{E}_1^0 = [\underbrace{1,\cdots,1}_{j^o},\underbrace{0,\cdots,0}_{j^s}],\ j^o = min\ \ k\ \ where\ \ \sum_{i=1,g_i\in G_j}^{k} P_i \geq L_j\ \ and\ j^o + j^s = M \\ \mathbf{Q}_1^0 = \left[S_1^0, D_1^0, S_1'^0, D_1'^0\right] = \left[\sum_{j=1}^{N}\max\left(\sum_{g_i\in G_j} P_i - L_j, 0\right), 0, \sum_{j=1}^{N}\max\left(\sum_{g_i\in G_j} P_i - L_j, 0\right), 0\right] \end{cases} \tag{8.11}$$

In Eq. (8.11), note the vector $\mathbf{E}_1^0$ represents whether the components are in online modes or warm standby modes. It depends on the component capacities and demand of each subsystem. j^o and j^s indicate the numbers of online components and standby components, respectively. For the illustrative example in Fig. 2, the values of the top node are presented in Eq. (8.12).

$$\begin{cases} \mathbf{B}_1^0 = (1, 1, 1, 1) \\ \mathbf{E}_1^0 = (1, 0, 1, 1) \\ \mathbf{Q}_1^0 = (12, 0, 12, 0) \end{cases} \tag{8.12}$$

Step 2: Construct the MDD for the first possible component failure.

Then, we can construct the MDD for the first possible component failure. Specifically, the first possible failure can occur to arbitrary component which results in (M + 1) possibilities. Therefore, there are (M + 1) branches for the first possible failure. The lth (l = 1,…, M) branch indicates the failure occurs to the lth component while the (M + 1)th branch represents no failure occurs. Figure 8.3 illustrates the MDD for the first failure.

In Fig. 8.3, the node values of the xth (x = 1,…, M) branches for the first failure can be presented in Eq. (8.13). The node values of the (M + 1)th branch is the same as that of the top node since no failure occurs for this branch.

$$\begin{cases} \mathbf{B}_x^1 = (\underbrace{1,\cdots,1}_{x-1}, 0, \underbrace{1,\cdots,1}_{M-x}) \\ \mathbf{Q}_x^1 = \left[S_x^1, D_x^1, S_x'^1, D_x'^1\right] \end{cases} \quad (x = 1,\cdots,M), \tag{8.13}$$

where we have

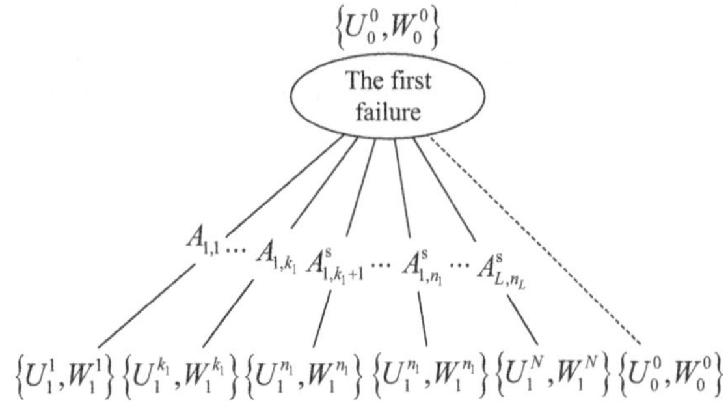

Fig. 8.3 MDD representation for the first failure

$$
\begin{cases}
S_x^1 = \max\left(S_1^0 - P_x, 0\right) = \sum_{j=1}^{N} \max\left(\sum_{g_i \in G_j, i \neq x} H_i - L_j, 0\right) \\
D_x^1 = \sum_{j=1}^{N} \max\left(L_j - \sum_{g_i \in G_j, i \neq x} H_i, 0\right) \\
S_x'^1 = S_x^1 - \min\left(S_x^1, D_x^1, C\right) = \\
\max\left(\begin{array}{l} 0, \sum_{j=1}^{N} \max\left(\sum_{g_i \in G_j, i \neq x} H_i - L_j, 0\right) - \\ \min\left(\sum_{j=1}^{N} \max\left(L_j - \sum_{g_i \in G_j, i \neq x} H_i, 0\right), C\right) \end{array}\right) \\
D_x'^1 = D_x^1 - \min\left(S_x^1, D_x^1, C\right) = \\
\max\left(\begin{array}{l} 0, \sum_{j=1}^{N} \max\left(L_j - \sum_{g_i \in G_j, i \neq x} H_i, 0\right) - \\ \min\left(\sum_{j=1}^{N} \max\left(\sum_{g_i \in G_j, i \neq x} H_i - L_j, 0\right), C\right) \end{array}\right)
\end{cases}
\tag{8.14}
$$

Note that $\mathbf{E}_x^1$ is determined by the performances of components and the demands in each subsystem. For example, if there exists performance deficiency after a component failure, the components in warm standby mode will be activated in a predesigned order to provide performance capacity to the subsystems with deficiency through common bus. The number of the activated components depends on the deficiency of the system. The warm standby components will be activated until the deficiency is made up.

Step 3: Construct the MDD for the second possible component failure.

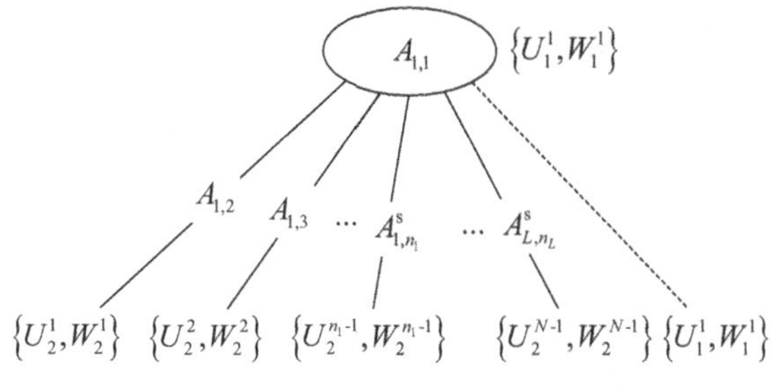

Fig. 8.4 MDD representation for the second failure (Taking the first failure occurs to the first component for example)

For the second failure, the MDD construction is based on the MDD for the first failure in Step 2. The terminal nodes in the MDD representation for the first failure are assumed to have M branches, where the xth ($x = 1, \cdots, M-1$) branch indicates a component failure while the Mth branch denotes no failure for this time. Taking the leftmost node in the MDD representation of the first failure for example as presented in Fig. 8.4, since the first failure occurs to the first component, for the second possible failure, there are M branches and the former (M-1) branches indicates the failures of the remaining (M-1) components while the Mth branch denotes no failure for this time. Then, the terminal values in the MDD are supposed to be updated which results from the second possible component failure. The updated values are formulated in Eq. (8.15).

$$\begin{cases} \mathbf{B}_x^2 = (0, \underbrace{1, \cdots, 1}_{x-1}, 0, \underbrace{1, \cdots, 1}_{M-x-1}) \\ \mathbf{Q}_x^2 = \left[S_x^2, D_x^2, S_x'^2, D_x'^2\right] \end{cases} \quad (x = 1, \cdots, M-1) \tag{8.15}$$

$$\begin{cases} S_x^2 = \max\left(S_1^1 - P_{x+1}, 0\right) = \sum_{j=1}^{N} \max\left(\sum_{g_i \in G_j, i \neq 1, i \neq x+1} H_i - L_j, 0\right) \\ D_x^2 = \sum_{j=1}^{N} \max\left(L_j - \sum_{g_i \in G_j, i \neq 1, i \neq x+1} H_i, 0\right) \\ S_x'^2 = S_x^2 - \min\left(S_x^2, D_x^2, C\right) = \\ \max\left(\begin{array}{l} 0, \sum_{j=1}^{N} \max\left(\sum_{g_i \in G_j, i \neq 1, i \neq x+1} H_i - L_j, 0\right) - \\ \min\left(\sum_{j=1}^{N} \max\left(L_j - \sum_{g_i \in G_j, i \neq 1, i \neq x+1} H_i, 0\right), C\right) \end{array}\right) \\ D_x'^2 = D_x^2 - \min\left(S_x^2, D_x^2, C\right) = \\ \max\left(\begin{array}{l} 0, \sum_{j=1}^{N} \max\left(L_j - \sum_{g_i \in G_j, i \neq 1, i \neq x+1} H_i, 0\right) - \\ \min\left(\sum_{j=1}^{N} \max\left(\sum_{g_i \in G_j, i \neq 1, i \neq x+1} H_i - L_j, 0\right), C\right) \end{array}\right) \end{cases} \tag{8.16}$$

Note that $\mathbf{E}_x^2$ is also determined by the performances of components and the demands in each subsystem which is similar to $\mathbf{E}_x^1$.

Step 4: Similarly, construct the MDD for the *wth* ($w = 3, \cdots, M$) possible component failure based on the MDD representation for the (*w*-1)*th* failure and update the node values according to the performances of system components and demands of subsystems. Figure 8.5 illustrates the system MDD for the proposed DBWSS.

Step 5: if the updated performance deficiency is not equal to zero, the MDD construction procedure is terminated because the system cannot satisfy the demand

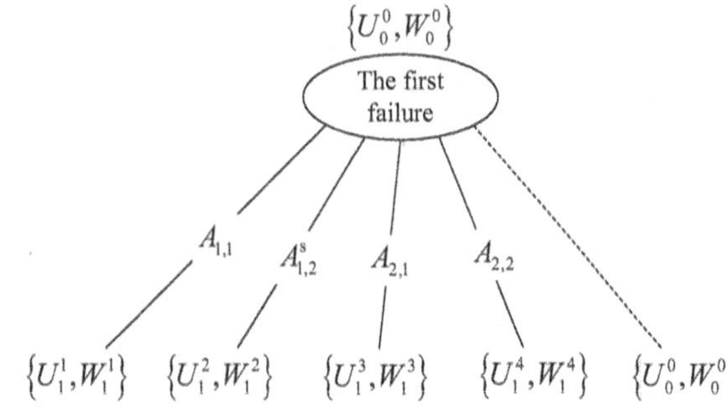

Fig. 8.5 MDD representation for the proposed DBWSS

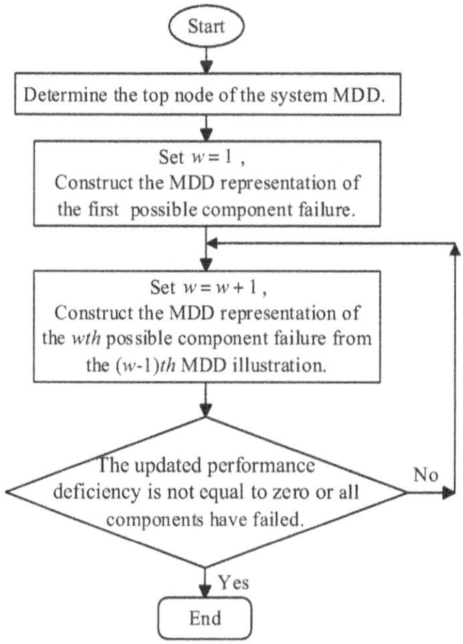

Fig. 8.6 The procedure for the MDD representation

of subsystems. Moreover, if there is no more component which can fail, the MDD construction is terminated as well. Therefore, the stopping criterion is the maximal number of failure $W = \max\ w(D_x^{w\prime} = 0)\ or\ W = M$.

Overall, according to the Steps above, the procedure for the construction of system MDD can be illustrated in Fig. 8.6. The system MDD illustration for the example in Fig. 8.2 is illustrated in Fig. 8.7.

8.3.2 System Reliability Evaluation Based on MDD

Taking the third path which indicates the failure occurs to the components A_3 and A_4 for example, this branch can be expressed as: component A_3 fails for the first time, then, component A_2 has to be activated from warm standby mode to provide performance for subsystem 2 through common bus; component A_4 fails for the second time, then, the performance is redistributed for compensating the performance deficiency of subsystem 2.

$$Path_1 = p_2 \int_0^t \int_{\tau_1}^t f_3(\tau_1) f_4(\tau_2) R_1(t) R_2^w(\tau_1) R_2^o(t - \tau_1) d\tau_1 d\tau_2, \tag{8.17}$$

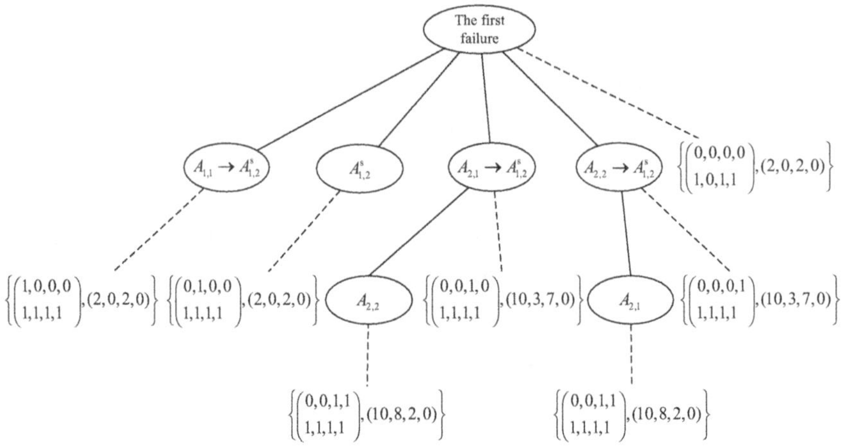

Fig. 8.7 The system MDD for the provided example in Fig. 8.2

where t is the mission time, $f_i(t)$ and $F_i(t)$ $(i = 1, \cdots, k)$ are the PDF and the CDF of the original online component, respectively. $f_i^o(t)$ and $f_i^w(t)(i = k+1, \cdots, M)$ are the PDF of the component failure in online mode or warm standby mode, respectively. Correspondingly, $F_i^o(t)$ and $F_i^w(t)(i = k+1, \cdots, M)$ are the CDF of the component failure in online mode or warm standby mode, respectively. $R_i(t) = 1 - F_i(t)$, $R_i^o(t) = 1 - F_i^o(t)$ and $R_i^w(t) = 1 - F_i^w(t)$ are the reliability functions of system components.τ_1 is an integral variable and locates in $(0, t)$ indicating the first failure time of the system. τ_2 is located in (τ_1, t) representing the failure time of the second system failure. p_i is the start failure probability of component A_i when activated from the warm standby mode to the online mode.

The reliability of the proposed system is the summation of the all the occurrence probabilities of the paths which result in a reliable system.

$$R(t) = \sum_b Path_b(t) \tag{8.18}$$

8.3.3 Complexity Analysis

The complexity of the proposed MDD construction mainly comes from the number of terminals in the system MDD. For the reliability assessment procedures, the complexity mainly lies in the integration for occurrence probabilities of paths in system MDD. Therefore, the computational complexity of the proposed method for the DBWSS with common bus performance sharing can be analyzed considering the terminals in the MDD construction and integration for occurrence probabilities, respectively.

According to the MDD construction procedures, for the first component failure, there are M branches indicating component failures. For each node in the second component failure, there are (M-1) branches denoting component failures. Therefore, there are $M(M\text{-}1)$ branches indicating component failures for M nodes. For the Wth component failure, there are $\prod_{w=1}^{W}(M-w+1)$ branches indicating component failures where $W = \max\ w(D_x^{w\prime} = 0)$ *or* $W = M$ indicats the maximal number of component failures leading a reliable system and w denotes the index for the number of component failure. Hence, we can obtain the number of total terminals in the system MDD is the summation of terminals of all possible component failures as formulated in Eq. (8.19).

$$\begin{aligned} & 1 + M + M(M-1) + \cdots + \prod_{w=1}^{W}(M-w+1) \\ & = 1 + \sum_{W}\prod_{w=1}^{W}(M-w+1) \end{aligned} \tag{8.19}$$

Specially, if the system MDD terminated due to no further component failure, this scenario would be the worst case with the most number of terminals. The total number of terminals of worst scenario is presented as Eq. (8.20).

$$\begin{aligned} & 1 + M + M(M-1) + \cdots + \prod_{w=1}^{M}(M-w+1) \\ & = 1 + \sum_{M}\prod_{w=1}^{M}(M-w+1) \end{aligned} \tag{8.20}$$

It indicates that the number of terminals of system MDD for the worst scenario is dependent on the number of components in the system. However, it should be noted that, generally, the number of terminals in the proposed algorithm is much smaller than Eq. (8.19) since the MDD construction is terminated when the demand is not satisfied. Moreover, the proposed MDD algorithm can be further improved by conducting unreliable paths if the number of unreliable paths is much less than that of reliable paths. In general, the proposed MDD method does not enumerate all the possible combinations, which can reduce the complexity of MDD conduction.

For example, in Fig. 8.7, the MDD construction is terminated after the second system failure. There are 7 terminals in the MDD leading to 7 reliable paths. However, there are 1 + 4 + 4*3 + 4*3*2 + 4*3*2*1 = 65 combinations for the illustrative example which is much larger than the number of paths for the proposed method.

For the computational effort of the integration for the occurrence probability, it is mainly from the multiple integration where several components are in failure. More details for the complexity of the multiple integration can be found in [7].

8.4 Numerical Studies

In this section, three cases are provided to verify the proposed method.

8.4.1 Case 1: Exponential Distribution

In order to validate the proposed MDD algorithm, Monte Carlo simulation technique is utilized to make comparison with the proposed method in this case.

Considering a DBWSS with performance sharing which is mentioned in model description, the mean-time-to-failure distributions for the components follow exponential distribution. The failure rates of the components in different modes are shown in Table 8.2. Four scenarios with different start failure probabilities and transmission capacities are analyzed as presented in Table 8.3. Moreover, in order to validate the effectiveness of the proposed MDD method, the proposed method is utilized in scenarios A, B, C and D while the Monte Carlo Simulation technique with 95% confidence intervals is applied to Scenario A' and B'. In this case, the computer programs were developed in MATLAB 2016b and were implanted on a laptop with a 2.00 GHz processor.

Figure 8.8 presents the time-varying reliabilities of the system for different scenarios in Case 1. It can be seen from Scenarios A and A', Scenarios B and B' that

Table 8.2 The failure rates of the components in different modes

Component /Failure rate	Online mode(/day)	Warm standby mode(/day)
A_1	1/100	–
A_2	1/150	1/300
A_3	1/150	–
A_4	1/100	1/300

Table 8.3 Scenarios with different transmission capacities and start failure probabilities

Scenario	A	B	A'	B'	C	D
Transmission capacity	10	5	10	5	10	5
Start failure probability	0	0	0	0	0.1	0.1
Utilizing method	MDD		Simulation		MDD	

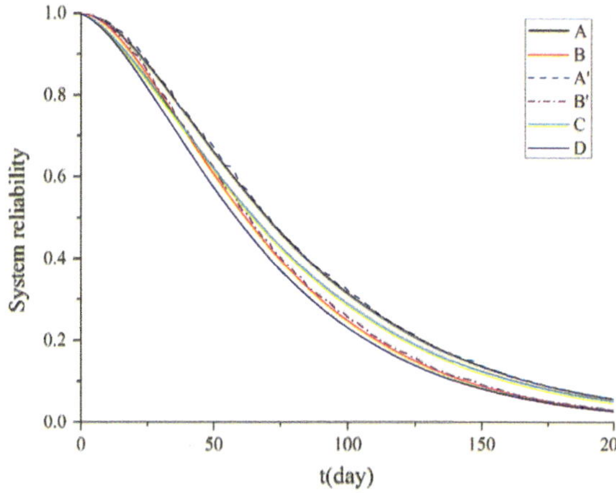

Fig. 8.8 System reliability of the DBWSS with common bus performance sharing for Case 1

the results of the proposed method approach to the results of Monte Carlo Simulation technique. Moreover, it is shown in Fig. 8.8 that with smaller transmission capacity, the system reliability is lower. It is because that some surplus performance might not be transmitted to the subsystems with performance deficiency due to the limitation of transmission capacity. Furthermore, the system reliability is lower with larger start failure probability, since the warm standby components have more possibilities to suffer from unsuccessful activations.

Especially, the numerical comparisons of reliability results of the proposed MDD algorithm and Monte Carlo Simulation for Scenario A and A' as well the relative differences are provided in Table 8.4. The relative differences indicate the minor differences between the results of the proposed method and Monte Carlo Simulation technique.

The computation time for different scenarios are presented in Table 8.5 which indicates that the computation time of the proposed method is much shorter than that of the Monte Carlo simulation. The computation time of Scenario A is a little longer than that of Scenario B due to larger number of paths in the system MDD.

Table 8.4 Comparisons of reliability results for Scenario A and A' in Case 1

Time (day)	Scenario A (MDD)	Scenario A' (Simulation)	Relative differences of different methods (%)
20	0.9137	0.9193	−0.61
50	0.6603	0.6735	−1.96
100	0.3141	0.3258	−3.59
200	0.0567	0.0573	−1.05

Table 8.5 Computation time for different scenarios in Case 1

Scenario	Computation time(s)
A	10.16
B	9.08
A'	360.67
B'	384.52
C	10.25
D	9.29

8.4.2 Case 2: Weibull Distribution

In order to validate the application of the proposed method in non-exponential distribution for component failure, the system with common bus whose component failures follow Weibull distribution is considered in this case. The reason for choosing Weibull distribution is the wide application in the reliability analysis of engineering systems. The parameters of the Weibull distribution for the component failure are listed in Table 8.6. The systems with different start failure probabilities and transmission capacities are considered as Scenarios A, B, C and D in Table 8.3.

The time-varying reliability of the presented system with Weibull distribution is illustrated in Fig. 8.9. The MDD construction is the same as that of Case 1. The differences lie in the integration expressions for different paths. The results are

Table 8.6 The parameters of the Weibull distribution for the component failure

Component	Online mode(/day)		Warm standby mode(/day)	
	Scale parameter	Shape parameter	Scale parameter	Shape parameter
A1	200	2	–	–
A2	150	1.5	300	1.5
A3	150	2	–	–
A4	200	1.5	300	1.5

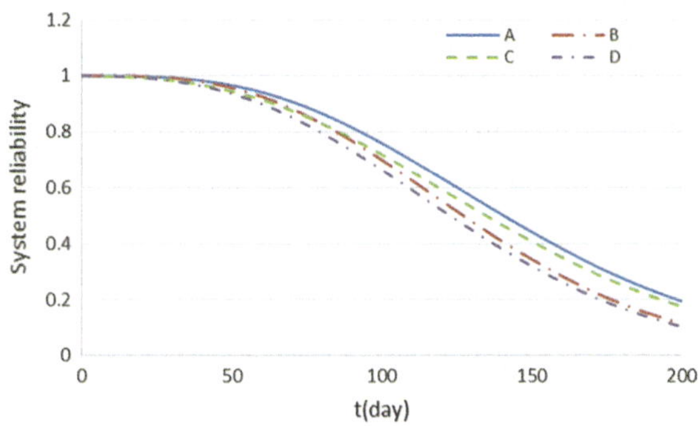

Fig. 8.9 The reliability of the DBWSS with Weibull distribution

consistent with those in Case 1 considering different transmission capacities and start failure probabilities.

8.4.3 Case 3: A DBWSS with Common Bus Consisting of 3 Subsystems

In this part, a DBWSS with common bus performance sharing system consists of 3 subsystem is provided. There are 9 components in the system whose characteristics including the performances and failure rates for its components are presented in Table 8.7. The reliability of systems for different scenarios with distinct start failure probabilities are provided.

For this case, there are 618 paths leading to a reliable system for the system MDD construction. Figure 8.10 presents the time-varying reliability of the DBWSS with common bus composed of 3 subsystems. It can be seen from Fig. 8.10 that with larger start failure probabilities, the system would become more unreliable.

Table 8.7 The characteristics of DBWSS with common bus in case 3

Subsystem (j)	Demand (L_j)	Component	Performance	Failure rate (/day)	
				Online	Warm standby
1	20	A_1, A_2	10	1/100	–
		A_3	10	1/150	1/300
2	30	A_4, A_5	15	1/150	–
		A_6	15	1/100	1/300
3	40	A_7, A_8	20	1/200	–
		A_9	20	1/150	1/400

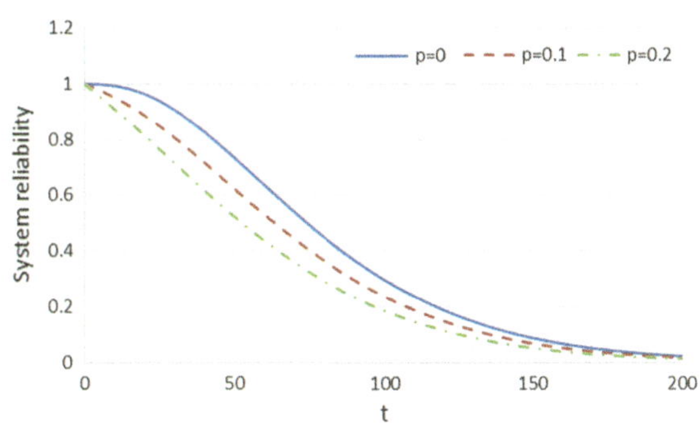

Fig. 8.10 The reliability of DBWSS with 3 subsystems in Case 3

8.5 Conclusions

With the wide application of redundancy techniques and common bus performance sharing systems, the model of DBWSS with common bus performance sharing is proposed. The imperfect switches of activating warm standby components are considered in the proposed model. Moreover, the MDD technique is developed to analyze the time-varying reliability of the proposed systems. The proposed method can be applied to systems with arbitrary time-to-failure distributions. The illustrative examples validate the proposed technique and the time-varying system reliabilities are presented. The common bus performance sharing systems with multi-state components and repairable components as well as the optimal activation sequence of warm standby components can be further studied.

References

1. Amari SV, Pham H, Misra RB. Reliability Characteristics of k-out-of-n Warm Standby Systems. *IEEE Transactions on Reliability* 2012; 61(4): 1007–1018.
2. Levitin G, Xing L, Dai Y. Cold vs. hot standby mission operation cost minimization for 1-out-of-N systems. *European Journal of Operational Research* 2014; 234(1): 155–162.
3. Xing L, Tannous O, Dugan JB. Reliability Analysis of Nonrepairable Cold-Standby Systems Using Sequential Binary Decision Diagrams. IEEE Transactions on Systems, Man, and Cybernetics - Part A: Systems and Humans 2012; 42(3): 715–726.
4. Tannous O, Xing L, J. B. Dugan. Reliability analysis of warm standby systems using sequential BDD. in 2011 Proceedings - Annual Reliability and Maintainability Symposium, Lake Buena Vista, FL, 2011, pp. 1–7
5. Zhang T, Xie M, Horigome M. Availability and reliability of k-out-of-(m+n): g warm standby systems. *Reliability Engineering & System Safety* 2006; 91(4): 381–387.
6. Zhai Q, Peng R, Xing L, Yang J. Reliability of demand-based warm standby systems subject to fault level coverage. *Applied Stochastic Models in Business and Industry* 2015; 31(3): 380–393.
7. Jia H, Ding Y, Peng R, Song Y. Reliability Evaluation for Demand-Based Warm Standby Systems Considering Degradation Process 2017; *IEEE Transactions on Reliability*, 66(3): 795–805.
8. Ding Y, Shao C, Yan J, Song Y, Zhang C, Guo C. Economical flexibility options for integrating fluctuating wind energy in power systems: The case of China 2018; *Applied Energy*, 228: 426–436.
9. Larsen EM, Ding Y, Li Y F, Zio E. Definitions of Generalized Multi-Performance Weighted Multi-State K^{-}-out-of-n System and its Reliability Evaluations. *Reliability Engineering & System Safety*, In Press. https://doi.org/https://doi.org/10.1016/j.ress.2017.06.009
10. Lisnianski A, Ding Y. Redundancy analysis for repairable multi-state system by using combined stochastic processes methods and universal generating function technique. Reliability Engineering & System Safety 2009; 94(11): 1788–1795.
11. Levitin G. Reliability of multi-state systems with common bus performance sharing. *IIE Transactions* 2011; 43: 518–524.
12. Xiao H, Peng R. Optimal allocation and maintenance of multi-state elements in series–parallel systems with common bus performance sharing. Computers & Industrial Engineering 2014; 72: 143–151.
13. Yu H, Yang J, Mo H. Reliability analysis of repairable multi-state system with common bus performance sharing. *Reliability Engineering & System Safety* 2014; 132: 90–96

14. Peng R, Liu H, Xie M. A Study of Reliability of Multi-State Systems with Two Performance Sharing Groups. *Quality and Reliability Engineering International* 2016; 32(7): 2623–2632
15. Xiao H, Shi D, Ding Y, Peng R. Optimal loading and protection of multi-state systems considering performance sharing mechanism. *Reliability Engineering & System Safety* 2016; 149: 88–95.
16. Yu H, Yang J, Zhao Yu. Reliability of nonrepairable phased-mission systems with common bus performance sharing. *Proceedings of the Institution of Mechanical Engineers, Part O: Journal of Risk and Reliability* 2018; online, DOI: https://doi.org/10.1177/1748006X18757074
17. Du S, Zeng Z, Cui L, Kang, R. Reliability analysis of Markov history-dependent repairable systems with neglected failures. *Reliability Engineering & System Safety* 2017; 159: 134–142.
18. Zeng Z, Zio, E. An integrated modeling framework for quantitative business continuity assessment. Process Safety and Environmental Protection 2017; 106: 76–88.
19. Zhai Q, Peng R, Xing L, Jun Y. Binary decision diagram-based reliability evaluation of k-out-of-(n+k) warm standby systems subject to fault-level coverage. *Proceedings of the Institution of Mechanical Engineers, Part O: Journal of Risk and Reliability* 2013; 227(5): 540–548.
20. Q Zhai L Xing R Peng J Yang 2015 Multi-Valued Decision Diagram-Based Reliability Analysis of k-out-of-n Cold Standby Systems Subject to Scheduled Backups IEEE Transactions on Reliability 64 4 1310–1324.
21. Mo Y. New Insights Into the BDD-Based Reliability Analysis of Phased-Mission Systems. *IEEE Transactions on Reliability* 2009; 58(4): 667–678.
22. Peng R, Zhai Q, Xing L, Yang J. Reliability of demand-based phased-mission systems subject to fault level coverage. *Reliability Engineering and System Safety* 2014; 121: 18-25.
23. Mo Y, Xing L, Amari SV. A Multiple-Valued Decision Diagram Based Method for Efficient Reliability Analysis of Non-Repairable Phased-Mission Systems. *IEEE Transactions on Reliability* 2014; 63(1): 320-330.
24. Xing L, Dai Y. A new decision-diagram-based method for efficient analysis on multistate systems. IEEE Transactions on Dependable and Secure Computing 2009; 6(3): 161-174.
25. Bryant RE. Graph-Based Algorithms for Boolean Function Manipulation. *IEEE Transactions on Computers* 1986; C-35(8): 677–691.
26. Wang ZQ, Wang W, Hu CH, Si XS. A prognostic-information-based order-replacement policy for a non-repairable critical system in service. IEEE Transactions on Reliability 2015; 64(2): 721-735

Chapter 9
Reliability of Warm Standby Systems with Phased-Mission Requirement

In the preceding chapters, we analyze the system reliability under fixed system demand. In reality, the demand can vary with time. In particular, the system mission may involve several tasks, and each task can have different demand for the system. For instance, a twin-engine aircraft flight involves taxi, take-off, ascent, level-flight, descent and landing phases. While a single engine can ensure the function of the taxi phase whereas both engines are required during the take-off phase. During each phase, the system has to accomplish a specified task and may suffer different environment or stress conditions. Thus, the system configuration, success criteria, reliability requirements, and component behavior may vary from phase to phase[1, 2]. Such cases are common in many industrial applications, e.g., the aviation, the nuclear industry and the telecommunication industry[3, 4]. The systems subject to multi-phase mission demand are called Phased-Mission System (PMS) [5, 6].

Because different phases have different requirements on the system configuration and operation, and the operating environments may also vary during the mission, the system would experience different stress levels, system success criteria and component failure behavior across the mission. In addition, the state of one component at the end of one phase is identical to its state at the beginning of the next phase, which inherently introduces inter-phase dependence[7]. Therefore, reliability modeling of PMSs is more challenging than single-phased systems. Xing and Amari [8] reviewed the studies on the system reliability modeling of PMS, and classified the existing studied into two categories: the analytical methods and the simulation methods. The simulation methods are flexible and can be easily modified to adapt to different situations. However, they are computationally inefficient and require considerable numbers of runs to evaluate the system reliability. The analytical methods can explicitly reflect the system structure and the results might be used in further applications, e.g. optimization of the system design. The analytical methods can be further classified into state-space oriented models [9–11], combinatorial methods [2, 12–14] and a phase modular solution [15] that combines the former two methods. The state-space oriented models (in particular Markov or Petri nets based methods) could handle both static and dynamic PMS. However they are difficult to be applied to large-scale

R. Peng et al., *Reliability Modelling and Optimization of Warm Standby Systems*,
https://doi.org/10.1007/978-981-16-1792-8_9

systems due to the well-known state-space explosion problem. The combinatorial methods exploit the Boolean algebra and decision diagrams to reduce the computational complexity, which makes them applicable to handle larger-scale systems. For example, Peng et al. (2014) proposed an MDD-based method for reliability analysis of a parallel PMS, and Peng et al. (2016) proposed a universal generating function based method for the reliability evaluation of series–parallel PMS.

Because of the complicated mission requirements for PMS, PMS is often designed with considerable level of redundancy to ensure its reliability. The warm standby design can balance the energy comsumption and the restoration time, which is suitable for PMS. In this chapter, we extend the MDD method to model the system reliability of a PMS with warm standby design.

9.1 System Description

Consider a non-repairable system with n statistically independent components $A_1, \ldots, A_n$ working in parallel. The system has to complete a mission with m consecutive non-overlapping phases. The jth phase has a predetermined duration $\tau_j, j = 1, \ldots, M$. Denote $T_j = \sum_{l=1}^{j} \tau_j$. Component A_i may fail in any of the m phases or survive the mission.

Each component has a nominal capacity $w_{i,j}$ in phase j when it is in normal state. Here, $w_{i,j}$ can vary with j to account for the performance dependence of A_i on environments, working conditions, etc. The system capacity equals to the sum of the capacities of all the working components, as in preceding chapters. The system capacity has to meet a predetermined mission demand d_j in phase j. A practical example of such systems is the power system in a region, which consists of multiple power plants A_i with variable capacity $w_{i,j}$ in different phases. The system capacity has to meet the power demand that may also vary with time. Denote $D = (d_1, \ldots, d_m)$. The mission succeeds if and only if the demand is satisfied in all the phases.

Assume that the lifetime distribution of a component is dependent on the working state of the system, and irrelevant with the phase the component is in. More specifically, we assume that the lifetime of component A_i follows an arbitrary distribution $F_i^s(t)$ in the warm standby state, and $F_i^o(t)$ in the normal working state. At the beginning of the phsed-mission, we assume that components $A_1, \ldots, A_k$ are at the normal working state, while the other $(n - k)$ components are at the warm standby state. Whenever a component failure occurs in a phase and leads to a decrease of system capacity, some warm standby components have to be switched on to ensure that the system capacity can satisfy the mission demand in that phase. The warm standby components are switched according to the index of the component in an ascent order. That is, A_{k+1} will be the first warm standby component to be switched on when needed, and then A_{k+2},A_{k+3}, until A_n.

The system is non-repairable in a failed component in some phase will stay in the failed state in the remaining mission.

9.2 Construction of System-Level Decision Diagram

We would analyze the system reliability based on MDD following a similar manner as in Chap. 4. The key in the reliability assessment of warm standby system is the enumeration of failure sequences during the mission. To accomplish this, we shall construct MDD to represent the possible failure sequences for the system. Based on the MDD, we can determine the occurrence probabilities for the edges in the MDD and calculate the system reliability.

As for the single-phase warm standby system, we consider the following general failure sequence for the PMS during the mission:

The first component failure occurs in phase j_1 and the failed component is A_{i_1}, the second component failure occurs in phase j_2 and the failed component is A_{i_2}, ..., and the rth component failure occurs in phase j_r and the failed component is A_{i_r}.

It can be verified that all the failure sequences in the mission can be covered by the above general description. Clearly, we should have $r \le n$ and $j_1 \le j_2 \le \ldots \le j_r$. To represent such a sequence, we can model each component failure in the sequence using an MDD as follows.

First cconsider the first component failure. The first failed component can be any one of the n components and it can fail in any of the m phases. Therefore, there are totally nm possible outcomes for the first failure. To enumerate all these possible cases, we build an MDD with $(nm + 1)$ branches, as shown in Fig. 9.1. In this MDD, the nm branches from left ro right represent the nm possible component failure scenarios in m phases, where the $((i - 1)m + j)$ th branch represents that the failed component is A_i and it fails at phase j. The terminal value of this branch is a $2 \times m$ matrix $M_{i,j}$. The first and second row of $M_{i,j}$ represent the the actual capacity loss and the available capacity loss in the m phases due to the failure of A_i in phase j, respectively. For the primary working components A_i, $1 \le i \le k$, if it fails at phase j, it will remain in the failed state for the remaining phases and the system will suffer a loss of capacity that should have contributed by A_i in the following phases due to

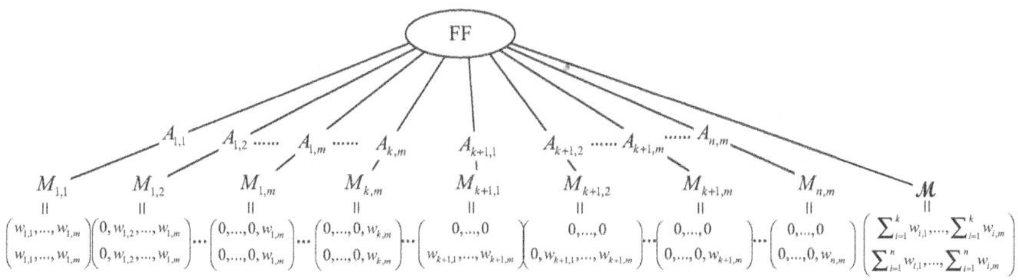

Fig. 9.1 MDD for the first component failure in the warm standby system with phased-mission

the failure of A_i. Because A_i is a primary working component, its failure will reduce the actual capacity as well as the available capacity for the system. Therefore, the terminal value for the $((i-1)m+j)$ th branch for $1 \le i \le k$ is

$$M_{i,j} = \begin{pmatrix} 0, \ldots, 0, w_{i,j}, \ldots, w_{i,n} \\ 0, \ldots, 0, w_{i,j}, \ldots, w_{i,n} \end{pmatrix}, i \le k.$$

For example, the terminal for the leftmost branch is

$$M_{1,1} = \begin{pmatrix} w_{1,1}, \ldots, w_{1,n} \\ w_{1,1}, \ldots, w_{1,n} \end{pmatrix},$$

representing that if the first component failure is that A_1 fails in phase 1, then the actual system capacity and the available system capacity in every phase will suffer a loss from A_1.

For warm standby component $A_i, k+1 \le i \le n$, if it fails in phase j, then the available system capacity in the following phases will decrease due to its failure. However, because the warm standby component has not provided any capacity in the standby state, it failure will not decrease the actual system capacity. Therefore, the terminal value of the $((i-1)m+j)$th for $k+1 \le i \le n$ is:

$$M_{i,j} = \begin{pmatrix} 0, \ldots, 0 \\ 0, \ldots, 0, w_{i,j}, \ldots, w_{i,n} \end{pmatrix}, i \ge k+1.$$

Thus, the nm branches with their terminal values completely represent all the possible scenarios for the first component failure in the PMS.

To complete the MDD, we add an additional branch in the rightmost to represent that the first component failure does not occur in the mission. In other words, there is no component failure throughout the whole mission. The terminal value for this branch is also a $2 \times m$ matrix, $\mathcal{M}$. Different from other branches, the first and second row of this matrix represent the actual system capacity and the available system capacity in the m phases given that the case represented by this branch happens, i.e., the first component failure does not occur. More specifically, the j th element of the first row is $\sum_{i=1}^{k} w_{i,j}$, which is equal to the sum of the capacities of the primary working components. This indicates that the actual system capacity is $\sum_{i=1}^{k} w_{i,j}$ if there is no component failure throughout the mission. Similarly, the j th element of the second row is $\sum_{i=1}^{n} w_{i,j}$, representing that the avaliable system capacity in phase j is $\sum_{i=1}^{k} w_{i,j}$ if there is no component failure throughout the mission.

Note that we would use the MDD to represent the possible scenarios that lead to system success, and will not consider the scenarios that lead to system failure. Therefore, we assume that the nm scenarios for the first component failure in Fig. 9.1

will not lead to system failure. In other words, the system may still meet the demand in the m phases even the first failure occurs:

$$\min\big(\mathcal{M}(2,:) - M_{i,j}(2,:) - D\big) \geq 0,$$

where $\mathcal{M}(2,:)$ and $M_{i,j}(2,:)$ represent the second rows of $\mathcal{M}$ and $M_{i,j}$, respectively.

Recall that we build the MDD for the second component failure based on the MDD for the first component failure using a four-step procedure for the single-phase warm standby system. For wamr standby PMS, we can use a similar procedure to build the MDD for the second component failure. For the $((i-1)m+j)$ th branch, we can build the MDD for the second component failure with the following steps given that first component failure corresponding to the $((i-1)m+j)$ th branch occurs.

Step-1: Copy the MDD for the first failure and change the node notation from "FF" to "$A_i, \big(t_0, t_{j,1}\big]$", representing that component A_i starts working from $t_0 = 0$ and fails at $t_{j,1}$, where the subscript of $t_{j,1}$ represents the first failure in phase j.

Step-2: Update the terminal of the rightmost branch to $\mathcal{M} = \mathcal{M} - M_{i,j}$. Note that the rightmost branch always denotes the complementary event of the union of the events represented by the left branches. More specifically, the rightmost branch represents the event that the system does not experience the second failure throughout the mission, which is equivalent to that the system only experiences one failure during the whole mission. This particular failure is just the one occurs in phase j and the failed component is A_i, i.e., the event corresponding to the $((i-1)m+j)$ th branch of the MDD for the first failure. This failure leads to a decrease of the actual system capacity and the available system capacity, and the reduced system capacity is represented by $\mathcal{M} - M_{i,j}$.

Step-3: Switch on the warm standby components. Because the failure of component A_i in phase j leads to a decrease of the system capacity, the actual system capacity at some phases in the mission may be lower than the demand. Because we have assumed that the failure of component A_i will not result in a system failure, which means that the available system capacity can still fulfil the demand in all the phases, some warm standby components can be switched on at some phases to ensure the actual system capacity in all the phases can meet the demand. We consider two possible cases that may occur after the failure of A_i in phase j: (1) the actual system capacity immediately decreases to a level below the demand for phase j, and (2) the actual system capacity immediately decreases, but it can still meet the demand at the moment until some phase l, $l > j$. In the first case, the warm standby components should be switched on at once to pull up the actual system capacity. In the second case, on the contrary, the warm standby component needs not to be switched on until phase l. Consider the first case. We first identify the $(m-j+1)$ branches corresponding to that A_{k+1} fails in phase s for $s = j, \ldots, m$, and replace the first row of the terminal matrix of each branch by the second row of the corresponding matrix. In other words, we change the terminal matrix as follows:

$$\begin{pmatrix} 0,\dots,0 \\ 0,\dots,0,w_{k+1,s},\dots,w_{k+1,m} \end{pmatrix} \rightarrow \begin{pmatrix} 0,\dots,0,w_{k+1,s},\dots,w_{k+1,m} \\ 0,\dots,0,w_{k+1,s},\dots,w_{k+1,m} \end{pmatrix}, s = j,\dots,m.$$

Recall that the first and the second row of the matrix represent the actual and the available capacity loss due to the failure of component A_{k+1} in phase s. Because A_{k+1} is in the warm standby state, its failure will not result in any actual capacity loss throught the mission, but will lead to the decrease of the available capacity from phase j on. Now, we replace the first row of this terminal matrix by its second row, to represent that the component is switched to the normal working state and its failure after switching will result in a loss to both the actual capacity and the available capacity. Meanwhile, the terminal of the rightmost branch is updated as follows to represent an increase of the actual system capacity from phase j on:

$$\mathcal{M}(1,s) = \mathcal{M}(1,s) + w_{k+1,s}, s = j,\dots,m,$$

where $\mathcal{M}(1,s)$ represents the s th element of the first row of $\mathcal{M}$. The actual system capacity is increased after the switching of component A_{k+1} in phase j. If the actual system capacity in phase j still cannot satisfy the demand, i.e., $\mathcal{M}(1,j) < d_j$, then we should further switch A_{k+2} on. Accordingly, the terminal matrice of these branches corresponding to that A_{k+1} fails in phase s for $s = j,\dots,m$ and the terminal matrix of the rightmost branch need to be updated similarly. The warm standby components are switched on sequentially, until the demand in phase j can be met. Suppose that there are r warm standby components $A_{k+1},\dots,A_{k+r}$ are switched on in total. To explicitly represent this action, we change the notation for the node to $A_i \rightarrow (A_{k+1},\dots,A_{k+r}), \left(t_0, t_{j,1}\right]$, with $A_i \rightarrow (A_{k+1},\dots,A_{k+r})$ representing that component A_i fails in phase j and components $A_{k+1},\dots,A_{k+r}$ are switched to the normal working state. Meanwhile, we supplement the time indicator $t_{j,1}$ to the terminals of the branches representing that $A_{k+1},\dots,A_{k+r}$ fail in phase $s = j,\dots,m$ to explicitly record the switching time of these components. More specifically, we change these terminals from $M_{k+1,s},\dots,M_{k+r,s}$ to $\left(M_{k+1,s}, t_{j,1}\right),\dots,\left(M_{k+r,s}, t_{j,1}\right)$.

Step-4: Remove unnecessary branches. For the obtained MDD, we shall delete the following branches:

(1) The m branches associated with component A_i. Because component A_i has failed in phase j as the first failure, it should not be considered in the modeling of the second failure.

(2) The branches representing component failures before phase j. Conditional on that the first failure occurs in phase j, all the failures that occurs before phase j should not be considered in the modeling of the second failure.

(3) Consider an arbitray branch among the remaining branches. Denote the terminal matrix by M and calculate $\mathcal{M}(2,s) - M(2,s)$. If $\min\{\mathcal{M}(2,:) - M(2,:) - D\} < 0$, it means that the available system capacity cannot meet the demand in some phases if the failure corresponding to this branch occurs. Delete this branch.

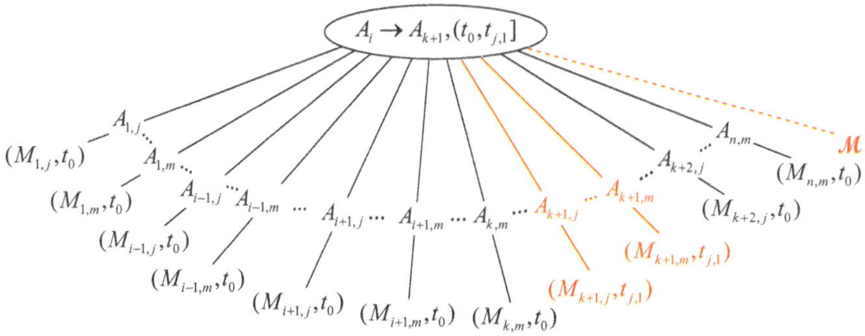

Fig. 9.2 MDD for the second failure given that the first failure is component A_i in phase j

After the above 4-step modification, we can obtain the MDD for the second failure given that the first failure is component A_i in phase j. For example, Fig. 9.2 illustrates the MDD for the second failure given that the first failure is A_i in phase j and the warm standby component A_{k+1} is switched to the normal working state. Note that the branches associated with A_i and the branches representing component failures before phase j are removed (we assume that the failures indicated by the remaining branches will not lead to the system failure). The terminal matrices of the branches of A_{k+1} and the rightmost branch are updated:

$$\tilde{M}(k+1, s) = \begin{pmatrix} 0, \ldots, 0, w_{k+1,s}, \ldots, w_{k+1,m} \\ 0, \ldots, 0, w_{k+1,s}, \ldots, w_{k+1,m} \end{pmatrix}, s = j, \ldots, m,$$

$$\mathcal{M}\widetilde{\mathcal{M}} = \begin{pmatrix} \sum_{l=1, l\neq i}^{k} w_{l,1}, \ldots, \sum_{l=1, l\neq i}^{k} w_{l,j-1}, w_{k+1,j} + \sum_{l=1, l\neq i}^{k} w_{l,j}, \ldots, w_{k+1,m} + \sum_{l=1, l\neq i}^{k} w_{l,m} \\ \sum_{l=1, l\neq i}^{n} w_{l,1}, \ldots, \sum_{l=1, l\neq i}^{n} w_{l,m} \end{pmatrix}.$$

In addition, we have also supplemented the time indicator to each of the terminals, where all the branches associated with the primary working components have the time indicator $t_0 = 0$ and the branches associated with the switched warm standby component A_{k+1} have the time indicator $t_{j,1}$. We temporarily assign the time indicator t_0 to the branches associated with all the unswitched wamr standby components.

Similarly, the MDD for the third component failure can be obtained based on the MDD for the second failure. Generally, the MDD for the l th component failure can be obtained from modifying the MDD for the $(l-1)$ th failure. During the modification, one should be careful wth the working time interval. For the l th failure after a branch of the MDD for the $(l-1)$ th failure, the intial working time indicator is just the one given in the terminal of the corresponding branch of the MDD for the $(l-1)$ th failure. For example, if we consider the MDD for the third failure after the leftmost branch of the MDD in Fig. 9.2, then the node notation for this MDD should have the following form:

$$A_1, (t_0, \sim],$$

where " ~ " represent the failure time (or mission termination time) indicator to determined. On the other hand, if we consider the MDD for the third failure after the branch marked by $A_{k+1,j}$ in Fig. 9.2, then the node notation of this MDD should have the following form

$$A_{k+1}, \left(t_{j,1}, \sim\right].$$

In other words, the left end of the working time interval is determined from the terminal value of the branch that this MDD is built after.

The right end of the working time interval, i.e., the failure time (or mission termination time) of the particular component, can be determined as follows. If the phase j_l in which the l th failure occurs is the same as the phase j_{l-1} in which the $(l-1)$ th failure occurs, then the failure time indicator is denoted by $t_{j_{l-1},r_{l-1}+1}$, where $t_{j_{l-1},r_{l-1}}$ denotes the failure time of the $(l-1)$ th failure and this failure is the r_{l-1} failure in phase j_{l-1}. If the phase j_l is after the phase j_{l-1}, i.e., $j_{l-1} < j_l$, then the failure time indicator for the l th failure is denoted by $t_{j_l,1}$, representing that the l th failure is the first failure in phase j_l. Consider the MDD for the third failure after the leftmost branch in Fig. 9.2, the node notation should have the following form:

$$A_1, \left(t_0, t_{j,2}\right].$$

This is because the first failure is component A_i in phase j, and if the second failure is component A_1 in phase j, its failure time should be denoted as $t_{j,2}$, indicating the second failure in phase j. On the contrary, if we build the MDD for the third failure after the second leftmost branch in Fig. 9.2, then the node notation should have the following form

$$A_1, \left(t_0, t_{j+1,1}\right].$$

This indicates that the second failure is component A_1 in phase $j+1$, and it is the first failure in phase $j+1$, which is denoted by $t_{j+1,1}$.

In the above 4-step MDD building procedure, we mentioned that there are two possible cases corresponding to a component failure in phase j: 1) the actual system capacity is reduced below the demand immediately, and the warm standby component should be switched on at once; and 2) the actual system capacity drops but the demand can still be met at phase j until some furture phase if no warm standby component is switched on. For the second case, we add the following substeps in Step-1:

Step-1.1: Assume the current failure occurs at phase j_l.

Step-1.2: Denote $S_d = \{j | 1 \le j \le m, \mathcal{M}_{1,j} - d_j < 0\}$. If S_d is nonempty and $\min\{S_d\} \le j_l$, it indicates that the demand cannot be satisfied at phase $\min\{S_d\}$ before this failure. Therefore, the available warm standby components are switched on at phase $\min\{S_d\}$ in order until the demand in phase $\min\{S_d\}$ is met. For each switched component, the terminal values for the corresponding branches and the rightmost branch are updated following Step-3. In addition, the time indicator is substituted by $T_{\min\{S_d\}-1}$ to indicate that the warm standby component is switched to the normal

Table 9.1 System configurations for a three component system with a 2-phase mission

Configuration		Demand
Component	Capacity in two phases	
A_1	(1, 2)	$d_1 = 3$
A_2	(2, 1)	$d_2 = 2$
A_3	(3, 2)	

working state at the beginning of phase $\min\{S_d\}$. If S_d is empty or $\min\{S_d\} > j_s$, then go to Step-2.

Step-1.3: Go to Step-1.2.

By incorporating these three substeps into the 4-step procedure, we complete the MDD building procedure. The system MDD is obtained as a byproduct of this building procedure by simply adding the MDD for the lth failure to the branches of the MDD for the $(l-1)$th failure.

To illustrate this process, we consider the example three-component warm standby system in Sect. 4.3. Suppose that the system has to complete another task after the task described in Sect. 4.3. The demand of the second task is $d_2 = 2$, and other system configurations are listed in Table 9.1.

According to the aforementioned procedure, the system MDD can be constructed as in Fig. 9.3. To save space, we omit the time indicator in the terminals. The time indicators in the MDD construction need some attention. For example, for the middle path

$$\text{FF} \rightarrow \left\{A_2 \rightarrow A_3, \left(t_0, t_{1,1}\right]\right\} \rightarrow \left\{A_3, \left(t_{1,1}, t_{2,1}\right]\right\} \rightarrow \begin{pmatrix} 4, 2 \\ 4, 2 \end{pmatrix},$$

it represents that the first failure is component A_2 in phase 1, and its failure triggers the switch of component A_3 from the warm standby state to the normal working state; the second failure is A_3 in phase 2, and the initial working time of A_3 (at the normal working state) is just its switching time $t_{1,1}$.

Every path in the MDD corresponds to a possible failure sequence during the mission. Every path is ended with a "rightmost branch" of a certain MDD, and the terminal values of the path record the actual system capacity and the available system capacity in all the phases. Similar as for the single-phase mission, here we obtain the sequences that lead to a success mission. Therefore, the system reliability can be obtained as the sum of the occurrence probabilities of all these paths.

9.3 Evaluation of System Reliability

To obtain the system reliability, we should calculate the occurrence probability of each path in the MDD. The occurrence probability of a path in the MDD can

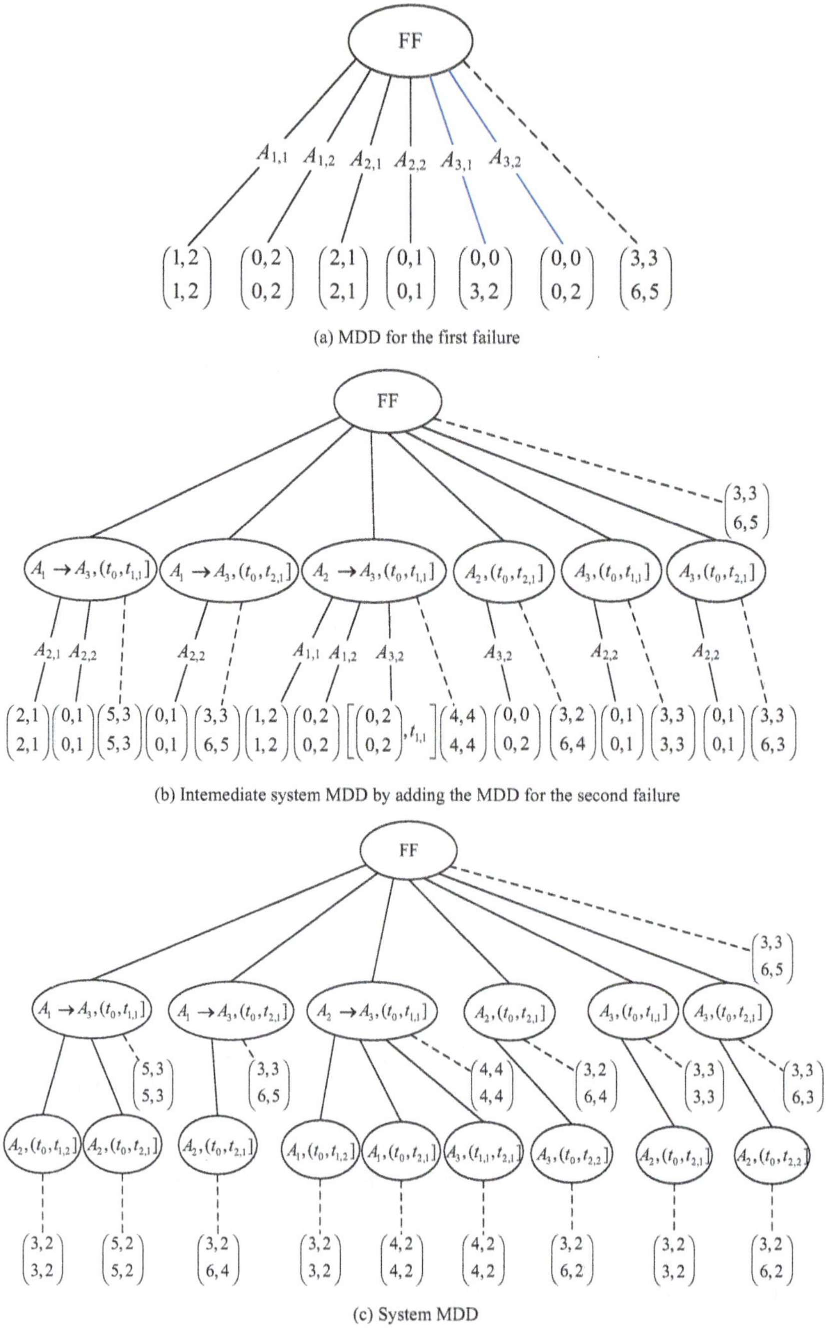

(a) MDD for the first failure

(b) Intemediate system MDD by adding the MDD for the second failure

(c) System MDD

Fig. 9.3 System MDD for the three-component warm standby system with a 2-phase mission

be obtained in a similar manner as for the system with single-phase missions, as described in Sect. 4.4. Consider a path in Fig. 9.3 for example:

$$\text{FF} \rightarrow \{A_2 \rightarrow A_3, (t_0, t_{1,1}]\} \rightarrow \{A_3, (t_{1,1}, t_{2,1}]\} \rightarrow \begin{pmatrix} 4, 2 \\ 4, 2 \end{pmatrix}.$$

This path corresponds to the scenario that the first failure is component A_2 in phase 1 and component A_3 is switched from the warm standby state to the online state, and the second failure is component A_3 in phase 2. The occurrence probability for this event is

$$\begin{aligned} \Pr\{Path\} &= R_1(T_2) \cdot \int_{t_0}^{T_1} f_2(t_{1,1}) R_3^{\mathrm{s}}(t_{1,1}) \left[F_3^{\mathrm{o}}(T_2 - t_{1,1}) - F_3^{\mathrm{o}}(T_1 - t_{1,1})\right] dt_{1,1} \\ &= R_0(T_2) \times \frac{\int_{t_0}^{T_1} R_3^{\mathrm{s}}(t_{1,1}) R_3^{\mathrm{o}}(T_1 - t_{1,1}) dF_2(t_{1,1})}{R_2(T_2) R_3^{\mathrm{s}}(T_2)} \times \frac{\int_{T_1}^{T_2} dF_3^{\mathrm{o}}(t_{2,1} - t_{1,1})}{R_3^{\mathrm{o}}(T_1 - t_{1,1})}, \end{aligned} \tag{9.1}$$

where T_1 and T_2 are the cumulative mission time at the end of phase 1 and phase 2, respectively. The occurrence probability in (9.1) is decomposed, where $R_0(T_2) = R_1(T_2)R_2(T_2)R_3^{\mathrm{s}}(T_2)$ corresponds to the probability that no failure occurs during the whole mission, the second term $\int_{t_0}^{T_1} R_3^{\mathrm{s}}(t_{1,1}) R_3^{\mathrm{o}}(T_1 - t_{1,1}) dF_2(t_{1,1}) / [R_2(T_2)R_3^{\mathrm{s}}(T_2)]$ corresponds to the occurrence probability of the edge leading to node $\{A_2 \rightarrow A_3, (t_0, t_{1,1}]\}$, and the third term $\int_{t_1}^{T} dF_2(t_2)/R_2(T)$ corresponds to the occurrence probability of the edge leading to node $\{A_3, (t_{1,1}, t_{2,1}]\}$.

In fact, the occurrence probability of every path can be decomposed in the same way. Generally, the nodes in the MDD have the following standard form

$$A_i \rightarrow (A_{i_1}, \ldots, A_{i_r}), (t_{j_1,s_1}, t_{j_2,s_2}], \tag{9.2}$$

where t_{j_1,s_1} and t_{j_2,s_2} represent the initial working time and the failure time of A_i, respectively, and the subscript "(j, s)" for $t_{j,s}$ indicates that the failure is in phase j and it is the sth failure in phase j. Clearly, $j_1 \le j_2$ and $s_1 < s_2$ for $j_1 = j_2$. This standard form denotes that component A_i starts working at t_{j_1,s_1} and fails at t_{j_2,s_2}, and warm standby components $A_{i_1}, \ldots, A_{i_r}$ are switched to the normal working state.

Denote $t_{j,0} = T_{j-1}$ and $t_{1,0} = t_0 = T_0$. The nodes can be classified into three types according to the initial working time: (1) $j_1 = 1, s_1 = 0$; (2) $j_1 > 1, s_1 = 0$; and (3) $s_1 > 0$. The three different types respectively correspond to the following three cases: (1) the component is an initial working component or a warm standby component, which starts working at the very beginning of the mission; (2) the component is a warm standby component and be switched on from the beginning of phase j_1; and (3) the component is a warm standby component and switched on immediately after a failure of the online working component. For the three different types of the nodes, the probability assignments to the edges leading to different nodes are:

$$\begin{cases} \dfrac{\int_{t_{j_2,s_2-1}}^{T_{j_2}} \prod_{v=1}^{r}\left[R_{i_v}^{s}(t_{j_2,s_2})R_{i_v}^{o}(T_m-t_{j_2,s_2})\right]dF_i^{s}(t_{j_2,s_2})}{R_i^{s}(T_m)\prod_{v=1}^{r}R_{i_v}^{s}(T_m)}, & j_1=1, s_1=0, \\ \dfrac{R_i^{s}(t_{j_1,0})}{R_i^{s}(T_m)}\cdot\dfrac{\int_{t_{j_2,s_2-1}}^{T_{j_2}} \prod_{v=1}^{r}\left[R_{i_v}^{s}(t_{j_2,s_2})R_{i_v}^{o}(T_m-t_{j_2,s_2})\right]dF_i^{o}(t_{j_2,s_2}-t_{j_1,0})}{\prod_{v=1}^{r}R_{i_v}^{s}(T_m)}, & j_1>1, s_1=0, \\ \dfrac{\int_{t_{j_2,s_2-1}}^{T_{j_2}} \prod_{v=1}^{r}\left[R_{i_v}^{s}(t_{j_2,s_2})R_{i_v}^{o}(T_m-t_{j_2,s_2})\right]dF_i^{o}(t_{j_2,s_2}-t_{j_1,s_1})}{\prod_{v=1}^{r}R_{i_v}^{s}(T_m)\times R_i^{o}(T_m-t_{j_1,s_1})}, & s_1\neq 0 \end{cases} \tag{9.3}$$

where the integration interval $\left(t_{j_2,s_2-1}, T_{j_2}\right)$ represents that component A_i fails after the failure of its preceding failure ($s_2 > 1$) or at any time during phase $j_2(s_2 = 1)$. Because the initial working components experience no warm standby state, we can simply assume that the lifetime distribution for an initial working component under the warm standby state is the same as that under the normal working state. In the above formulas, the first line corresponds to the case that the failed component is the initial working component or the warm standby component at the standby state, the occurrence probability is $dF_i^{s}\left(t_{j_2,s_2}\right)$. The second line corresponds to the case that the system can satisfy the mission demand in the preceding phase but the capacity is deficient after the phase transition. The warm standby is switched on at the beginning of a certain phase $t_{j_1,0}$, and fails at t_{j_2,s_2} with probability $dF_i^{o}\left(t_{j_2,s_2} - t_{j_1,0}\right)$. The third line corresponds to the case that a warm standby component is switched on at t_{j_1,s_1} to replace a failed component, and it fails at t_{j_2,s_2} with probability $dF_i^{o}\left(t_{j_2,s_2} - t_{j_1,s_1}\right)$.

The different formulas in (9.3) are used to characterize different scenarios of warm stanby components. If we let $j_1 = 1$ in the second formula in (9.3), we have

$$\frac{\int_{t_{j_2,s_2-1}}^{T_{j_2}} \prod_{v=1}^{r}\left[R_{i_v}^{s}(t_{j_2,s_2})R_{i_v}^{o}(T_m-t_{j_2,s_2})\right]dF_i^{o}(t_{j_2,s_2})}{R_i^{s}(T_m)\prod_{v=1}^{r}R_{i_v}^{s}(T_m)}. \tag{9.4}$$

If we let $j_1 = 1, s_1 = 0$ in the third formula in (9.3), we have

$$\frac{\int_{t_{j_2,s_2-1}}^{T_{j_2}} \prod_{v=1}^{r}\left[R_{i_v}^{s}(t_{j_2,s_2})R_{i_v}^{o}(T_m-t_{j_2,s_2})\right]dF_i^{o}(t_{j_2,s_2})}{R_i^{o}(T_m)\prod_{v=1}^{r}R_{i_v}^{s}(T_m)}. \tag{9.5}$$

The difference between (9.3) and (9.4) lies in that the denominator in (9.4) involves $R_i^{s}(T_m)$ while the denominator in (9.5) involves $R_i^{o}(T_m)$. The difference between (9.4) and (9.3) lies in that the integration variable in (9.3) is $dF_i^{s}\left(t_{j_2,s_2}\right)$ but the integration variable in (9.4) is $dF_i^{o}\left(t_{j_2,s_2}\right)$. For the initial working components, we have assumed that $R_i^{s}(\cdot) = R_i^{o}(\cdot)$, $F_i^{s}(\cdot) = F_i^{o}(\cdot)$, which indicates that the three formulas are equivalent for the initial working components.

According to the probability assignment rules, we can obtain all the probabilities associated with the edges in the MDD. We further assigne probability 1 to the edges

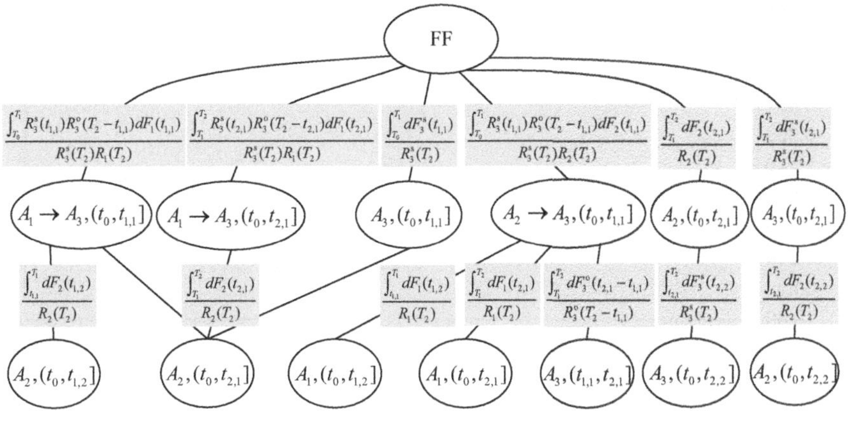

Fig. 9.4 Probabilities associated with each edge in the MDD

leading to the terminal nodes, and denote $R_0(T_m) = \prod_{i=1}^{k} R_i(T_m) \cdot \prod_{i=(k+1)}^{n} R_i^s(T_m)$. Then, the occurrence probability of each path can be obtained as the product of the probabilities of the edges in the path with $R_0(T_m)$. The system reliability can be obtained as the sum of the occurrence probabilities of all the paths. Figure 9.4 illustrates the probability assignment to all the edges in the example MDD in Fig. 9.3. To save space, we have delete all the edges that lead to the terminal nodes.

9.4 Numerical Example

Consider the aforementioned warm standby PMS with three components and 2 phases. Assume that the durations of the two phases are $\tau_1 = \tau_2 = 1$. Denote $T_0 = 0,\ T_1 = 1,\ T_2 = 2$. Assume that the lifetime distributions of the three components in the warm standby state are $F_i^s(t) = 1 - e^{-0.1t}$, and the lifetime distributions in the normal working state are $F_i^o(t) = 1 - e^{-0.2t}$.

From Fig. 9.4, it can be noted that the system success consists of 9 possible sequences that contains two component failures, 6 possible sequences that contains one component failure and one sequence that no failure occurs during the mission. Therefore, there are 16 possible scenerios that contributes to the system success. The occurrence probability of each path can be obtained directly from the probabilities associated with the edges in the path. For example, the occurrence probability for the leftmost path is

$$\Pr\{Path\} = R_0(T_2) \times \frac{\int_{T_0}^{T_1} R_3^s(t_{1,1}) R_3^o(T_2 - t_{1,1}) dF_1(t_{1,1})}{R_3^s(T_2) R_1(T_2)} \times \frac{\int_{t_{1,1}}^{T_1} dF_2(t_{1,2})}{R_2(T_2)} = 0.0114.$$

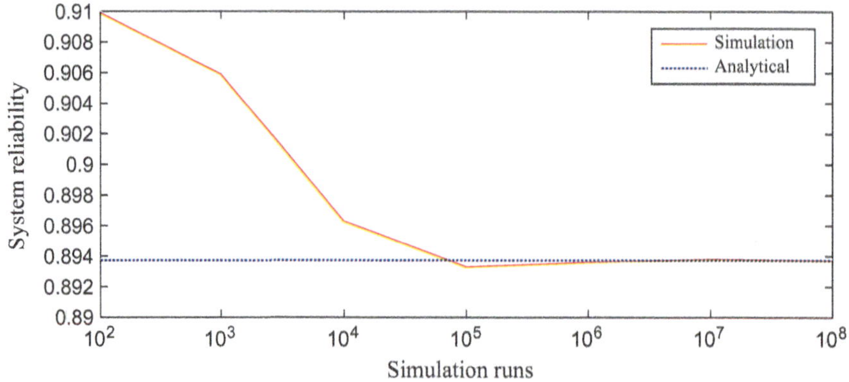

Fig. 9.5 System reliability from the MDD-based method versus the Monte Carlo method with different sample size

The system reliability is obtained as the occurrence probabilities of 16 paths in the MDD. For the example system, we can obtain the system reliability with the given parameter setting as 0.8938. Figure 9.5 shows the system reliability from the Monte Carlo method versus the MDD-based method. As can be noted, the result from the Monte Carlo method approaches the result of the MDD-based method, indicating the correctness of the MDD-based method.

References

1. Pedar A, Sarma V. Phased-mission analysis for evaluating the effectiveness of aerospace computing-systems [J]. IEEE Transactions on Reliability, 1981, 30: 429–437.
2. Zang X Y, Sun H R, Trivedi, K S. A BDD-based algorithm for reliability analysis of phased-mission systems [J]. IEEE Transactions on Reliability, 1999, 48(1): 50–60.
3. Peng R, Zhai Q, Xing L, Yang J. Reliability of demand-based phased-mission systems subject to fault level coverage [J]. Reliability Engineering & System Safety, 2014, 121: 18–25.
4. 许双伟, 武小悦. 高可靠多阶段任务系统可靠性仿真的高效方法[J]. 装备学院学报, 2012, 23(3): 69–74.
5. Mo YC. Variable ordering to improve BDD analysis of phased-mission systems with multimode failures [J]. IEEE Transactions on Reliability, 2009, 58(1): 53–57.
6. 莫毓昌, 杨孝宗, 崔刚, 刘宏伟. 一般阶段任务系统的任务可靠性分析[J]. 软件学报. 2007, 18(4): 1068–1076.
7. 孟礼, 武小悦. 基于BDD算法的航天测控系统任务可靠性建模与分析[J]. 装备学院学报, 2015 (5): 113–119.
8. Xing L, Amari S V. Reliability of phased-mission systems [M]//MISRA K B. Handbook of Performability Engineering, 2008, 349–368.
9. Smotherman M, Zemoudeh M. A non-homogeneous Markov model for phased-mission reliability analysis [J]. IEEE Transactions on Reliability, 1989, 38: 585–590.
10. Bondavalli A, Chiaradonna S, Di Giandomenico F, Mura I. Dependability modeling and evaluation of multiple-phased systems using DEEM [J]. IEEE Transactions on Reliability, 2004, 53: 509–522.

11. Chew S P, Dunnett S J, Andrews J D. Phased mission modelling of systems with maintenance-free operating periods using simulated Petri nets [J]. Reliability Engineering & System Safety, 2008, 93: 980–994.
12. Xing L, Dugan J B. Analysis of generalized phased-mission system reliability, performance, and sensitivity [J]. IEEE Transactions on Reliability, 2002, 51: 199–211.
13. Remenyte-Prescott R, Andrews JD, Chung P. An efficient phased mission reliability analysis for autonomous vehicles [J]. Reliability Engineering & System Safety, 2010, 95:226–235.
14. Tang Z, Dugan JB. BDD-based reliability analysis of phased-mission systems with multimode failures [J]. IEEE Transactions on Reliability, 2006, 55: 350–360.
15. Shrestha A, Xing L, Dai Y. Reliability analysis of multistate phased-mission systems with unordered and ordered states [J]. IEEE Transactions on Systems, Man and Cybernetics-Part A: Systems and Humans, 2011, 41: 625–636.
16. Ma Y, Trivedi K. An algorithm for reliability analysis of phased-mission systems [J]. Reliability Engineering & System Safety, 1999, 66:157–170.
17. Peng R, Zhai Q, Xing L, Yang J. Reliability analysis and optimal structure of series-parallel phased-mission systems subject to fault level coverage [J]. IIE Transactions, 2016, 48 (8): 736–746.

Chapter 10
Reliability of Warm Standby Systems with Complex Structure

10.1 Warm Standby of Complex Structure

The redundant systems with warm standby components studied in the previous chapters are all simple parallel structures. In practice, the systems may have more complex structures, such as series-parallel system, consecutive k-n system, linear sliding window system, etc.

Series-parallel system consists of several parallel subsystems in series [1, 2]. Examples of series-parallel system include energy system, nuclear power system, etc. There have been many studies on redundancy allocation in series-parallel systems, see [3–10]. Specifically, Peng et al. [11] studied the optimal structure of a series-parallel data transmission system. Xiao and Peng [12] studied the optimal structure of series-parallel system with performance sharing group.

Consecutive k-n system includes consecutive k-n: G system and consecutive k-n system: F system. There are n components in both systems. G system means that the system works only when there are at least k consecutive working components in the system, while F system means that the system fails when at least k consecutive components fail. Consecutive k-n system can also be divided into linear consecutive k-n system and circular consecutive k-n system. Examples of consecutive k-n systems include communication transmission systems and control systems [13–15]. A large number of literatures have studied consecutive k-n systems and other consecutively connected systems [16–19]. Shen and Cui [20] studied a sparse connected circular consecutive system. Daus and Beiu [21] studied the upper and lower bounds of the reliability of a linear consecutively connected system. Zhu et al. [22] studied the reliability importance of elements in two different consecutive k-n systems with components dependence. Peng et al. [18] studied the optimal component allocation problem in a consecutive system. Yu et al. [23] studied the reliability of a consecutive system with two failure modes.

Linear sliding window system is an extension of linear consecutive k-n system. The system is composed of n linearly arranged components. Only when the total performance of consecutive k components is not less than a specified demand D, can

R. Peng et al., *Reliability Modelling and Optimization of Warm Standby Systems*,
https://doi.org/10.1007/978-981-16-1792-8_10

the system work normally [24]. Examples of the system can be seen in communication system [25], computer system, etc. [26]. Some researchers have studied the reliability of sliding window system and its extended structure system [27–30]. Xiao et al. [31] studied the optimal components allocation problem in a sliding window system. Xiao et al. [32] studied the optimal load distribution problem of a sliding window system. Based on the linear sliding window system, there are many other extended structures, see references [28, 29, 33–35]. For the reliability research of the system with more complex topology, refer to the literature [36–42].

10.2 System Reliability

This section focuses on how to establish the reliability model of the system with warm standby components with various complex structures. The overall modeling idea is to decompose the system into several subsystems, and the reliability of the system can be expressed as a function of the reliability of each subsystem.

For example, a series-parallel system contains N subsystems, and each subsystem i has k_i work components and $n_i - k_i$ warm standby components. If the working component and warm standby component in each subsystem have the same nominal capacity, then each subsystem is a k-out-of-n system. The reliability R_i of each subsystem i can be calculated by using the method in Chap. 3. In this way, the reliability of the system can be expressed as

$$R_S = 1 - \prod_{i=1}^{N}(1 - R_i). \tag{10.1}$$

If the components in each subsystem have different nominal capabilities, the system capacity is the minimum value of each subsystem capability, and the external environment has a given demand for the system, then each subsystem is a demand-based warm standby system. For example, a series-parallel data transmission system may include several transmission subsystems, and each subsystem may contain several transmission channels. The data transmission speed of series parallel data transmission system is determined by the subsystem with the smallest transmission speed. Therefore, if the reliability of the series-parallel transmission system is defined as the probability that the data transmission speed of the system is greater than a given value in a certain period of time, the probability that the transmission speed of each subsystem is greater than the given value should be calculated. In order to obtain high system reliability, each subsystem can choose to set part of the transmission channel to the working state, and set the other channels to the warm standby state. In this case, the reliability of each subsystem can be calculated by using the method in Chap. 4, and then the system reliability can be obtained by using Eq. (10.1).

For the consecutive k-n system with warm standby components, we can use the same idea to study the reliability of the system. For example, a monitoring system may include n monitoring points, which are arranged in a straight line with a certain distance between them. If the monitoring equipment of k consecutive monitoring points cannot complete the monitoring of the corresponding monitoring area, it may cause the monitoring blind area in the key area and cause system failure. In order to ensure the high reliable operation of the system, a number of monitoring equipment can be placed at each monitoring point, some of which are in working state and some in warm standby state. As long as there are a certain number of monitoring equipment in each monitoring point, the monitoring task of the monitoring point can be completed. For this kind of system, each node is a warm standby subsystem, and the whole system is a consecutive k-n system with several nodes. In order to calculate the reliability of the system, we only need to calculate the reliability of the warm standby subsystem at each monitoring point, and then substitute it into the reliability calculation formula of the consecutive k-n system. Similarly, if the nominal capacity of each monitoring equipment at a monitoring point is the same, the method in Chap. 3 needs to be used; otherwise, the method in Chap. 4 needs to be used. In addition, if the failure of monitoring equipment cannot be detected in time, the problem of fault coverage should be considered. If each monitoring equipment has a certain probability of automatic alarm when it fails, then it conforms to the element level fault coverage model discussed in Chap. 5. Moreover, the switching failure between monitoring devices can also be considered by using the method in Chap. 5.

The sliding window system can also be extended to include warm standby components. For example, a heating system may have multiple heating subsystems, and these subsystems are arranged linearly. Each subsystem can contain several heaters, some of which are in working state and some in warm standby state. When the performance of several consecutive subsystems is less than a certain value, the heating in certain area will be insufficient, which will lead to system failure. Without losing generality, it can be assumed that the performance of each heater is a multi-valued random variable. Therefore, we can first use the method in Chap. 7 to get the multi-valued decision diagram of each subsystem to represent the possible states of each subsystem. By calculating the probability of each path, the performance distribution of each subsystem can be obtained. According to the performance distribution of each subsystem, the universal generating function method can be used to calculate the reliability of the system.

Similarly, for a more complex system, it can be decomposed into several subsystems with and without warm standby components, and then the reliability or performance distribution of each subsystem can be obtained step by step, and the reliability of the whole system can be calculated accordingly. Of course, the multi-valued decision diagram method based on failure sequence in Chap. 4 can also be used to directly consider every possible failure sequence in the system, and calculate the occurrence probability of each path, and then obtain the system reliability. By synthesizing all kinds of possible failure sequences, the performance distribution of the system can be obtained, and then the reliability of the system can be obtained.

In practical application, in addition to calculating the system reliability, we can also study the optimal redundancy configuration in the system, such as selecting the best number of different types of warm standby components for each subsystem. In the optimization, the reliability and cost of the system can be considered comprehensively to carry out multi-objective optimization. In a word, the reliability modeling method of the system with warm standby components proposed in this book can realize the reliability evaluation and optimization of various complex systems with warm standby components, so as to achieve the purpose of saving cost, safe production and improving profits.

References

1. Kolowrocki K, Kwiatuszewska-Sarnecka B. Reliability and risk analysis of large systems with ageing components [J]. Reliability Engineering & System Safety, 2008, 93 (12): 1821–1829.
2. Levitin G, Lisnianski A. A new approach to solving problems of multi-state system reliability optimization [J]. Quality and Reliability Engineering International, 2001, 17 (2): 93–104.
3. Agarwal M, Gupta R. Homogeneous redundancy optimization in multi-state series-parallel systems: A heuristic approach [J]. IIE Transactions, 2007, 39 (2): 277–289.
4. Hsieh T J, Yeh W C. Penalty guided bees search for redundancy allocation problems with a mix of components in series-parallel systems [J]. Computers and Industrial Engineering, 2012, 39 (11): 2688–2704.
5. Ramirez-Marquez JE, Coit DW. A heuristic for solving the redundancy allocation problem for multi-state series-parallel systems [J]. Reliability Engineering & System Safety, 2004, 83 (3):341–349.
6. Salmasnia A, Ameri E, Niaki S. A robust loss function approach for a multi-objective redundancy allocation problem [J]. Applied Mathematical Modelling, 2016, 40 (1): 635–645.
7. Soltani R, Safari J, Sadjadi S. Robust counterpart optimization for the redundancy allocation problem in series-parallel systems with component mixing under uncertainty [J]. Applied Mathematics and Computation, 2015, 271: 80–88.
8. Tian Z G, Zuo M J, Huang H Z. Reliability-redundancy allocation for multi-state series-parallel systems [J]. IEEE Transactions on Reliability, 2008, 57 (2): 303–310.
9. Tian Z G, Levitin G, Zuo M J. A joint reliability-redundancy optimization approach for multi-state series-parallel systems [J]. Reliability Engineering & System Safety, 2009, 94 (10): 1568–1576.
10. Zhou Y, Tian R, Sun Y, et al. An effective approach to reducing strategy space for maintenance optimisation of multistate series-parallel systems [J]. Reliability Engineering & System Safety, 2015, 138: 40–53.
11. Peng R, Mo HD, Xie M, Levitin G. Optimal structure of multi-state systems with multi-fault coverage [J]. Reliability Engineering & System Safety, 2013, 119: 18–25.
12. Xiao H, Peng R. Optimal allocation and maintenance of multi-state elements in series–parallel systems with common bus performance sharing [J]. Computers & Industrial Engineering, 2014, 72: 143–151.
13. Levitin G. Optimal allocation of multi-state transmitters in acyclic transmission network [J]. Reliability Engineering & System Safety, 2001, 75: 73–82.
14. Levitin G, Xing L, Dai Y. Optimal allocation of connecting elements in phase mission linear consecutively-connected systems [J]. IEEE Transactions on Reliability, 2013, 62 (3):618–627.
15. 何爱民,赵先,崔利荣,解伟娟. 线形可重叠的m-consecutive-k-out-of-n:F系统可靠性和单元重要度研究[J]. 兵工学报, 2009, 30 (z1): 135–138.
16. Kossow A, Preuss W. Reliability of linear consecutively-connected systems with multistate components [J]. IEEE Transactions on Reliability, 1995, 44 (3): 518 –522.
17. Levitin G. Optimal allocation of multi-state elements in linear consecutively connected systems with vulnerable nodes [J]. European Journal of Operational Research, 2003, 150 (2): 406–419.

18. Peng R, Xie M, Ng SH, Levitin G. Element maintenance and allocation for linear consecutively connected systems [J]. IIE Transactions, 2012, 44 (11): 964–973.
19. Zuo M J. Reliability of multistate consecutively-connected systems [J]. Reliability Engineering & System Safety, 1994, 44 (2): 173–176.
20. Shen J, Cui L. Reliability and birnbaum importance for sparsely connected circular consecutive-k systems [J]. IEEE Transactions on Reliability, 2015, 64 (4): 1140–1157.
21. Daus L, Beiu V. Lower and upper reliability bounds for consecutive-k-out-of-n:f systems [J]. IEEE Transactions on Reliability, 2015, 64 (3): 1128–1135.
22. Zhu, X, Boushaba M, Reghioua, M. Joint reliability importance in a consecutive-k-out-of-n:F system and an m-consecutive-k-out-of-n:F system for markov-dependent components [J]. IEEE Transactions on Reliability, 2015, 64 (2): 784–798.
23. Yu H, Yang J, Peng R, Zhao Y. Linear multi-state consecutively-connected systems constrained by m consecutive and n total gaps [J]. Reliability Engineering & Systems Safety, 2016, 150:35–43.
24. Levitin G. Linear multi-state sliding-window systems [J]. IEEE Transactions on Reliability, 2003, 52 (2): 263–269.
25. Li B, Blasch E. Two-way handshaking circular sequential k-out-of-n congestion system [J]. IEEE Transactions on Reliability, 2008, 57 (1):59–70.
26. 撒鹏飞, 赵敏. 微小卫星星务计算机系统模拟部件可靠性设计[J]. 系统工程与电子技术. 2006 (2):313–316.
27. Konak A, Kulturel-Konak S, Levitin G. Multi-objective optimization of linear multi-state multiple sliding window system [J]. Reliability Engineering & System Safety, 2012, 98 (1):24–34.
28. Levitin G, Ben-Haim H. Consecutive sliding window systems [J]. Reliability Engineering & System Safety, 2011, 96 (10): 1367–1374.
29. Levitin G, Dai Y. k-out-of-n sliding window systems [J]. IEEE Transactions on Systems Man and Cybernetics-Part A: Systems and Humans, 2012, 42 (3): 707–714.
30. Xiang Y, Levitin G, Dai Y. Optimal allocation of multistate components in consecutive sliding window systems [J]. IEEE Transactions on Reliability, 2013, 62 (1): 267–75.
31. Xiao H, Peng R, Levitin G. Optimal replacement and allocation of multi-state elements in k-within-m-from-r/n sliding window systems [J]. Applied Stochastic Models in Business and Industry, 2015, 32 (2): 184–198.
32. Xiao H, Peng R, Wang W, Zhao F. Optimal element loading for linear sliding window systems [J]. Proceedings of the Institution of Mechanical Engineers, Part O: Journal of Risk and Reliability, 2016, 230: 75–84.
33. Levitin G. Optimal allocation of elements in a linear multi-state sliding window system [J]. Reliability Engineering & System Safety, 2002, 76 (3): 245–254.
34. Levitin G. Uneven allocation of elements in linear multistate sliding window system [J]. European Journal of Operational Research, 2005, 163: 418–433.
35. 宋月, 刘三阳, 冯海林. 相邻k-out-of-n:F多状态可修系统的可靠性分析[J].系统工程与电子技术, 2006 (2): 310–312.
36. Lin Y K. A simple algorithm for reliability evaluation of a stochastic-flow network with node failure [J]. Computers & Operations Research, 2001, 28 (13): 1277–1285.
37. Lin Y K. System reliability evaluation for a multistate supply chain network with failure nodes using minimal paths [J]. IEEE Transactions on Reliability, 2009, 58 (1): 34–40.
38. Lin Y, Huang C, Yeh C. Assessment of system reliability for a stochastic-flow distribution network with the spoilage property [J]. International Journal of Systems Science, 2016, 47 (6): 1421–1432.
39. Yeh W. A fast algorithm for quickest path reliability evaluations in multi-state flow networks [J]. IEEE Transactions on Reliability, 2015, 64 (4): 1175–1184.

40. Yeh W. An improved sum-of-disjoint-products technique for symbolic multi-state flow network reliability [J]. IEEE Transactions on Reliability, 2015, 64 (4): 1185–1193.
41. Yeh W, Bae C, Huang C. A new cut-based algorithm for the multi-state flow network reliability problem [J]. Reliability Engineering & System Safety, 2015, 136: 1–7.
42. Zuo M J, Tian Z G, Huang H Z. An efficient method for reliability evaluation of multistate networks given all minimal path vectors [J]. IIE Transactions, 2007, 39 (8): 811–817.

The manufacturer's authorised representative in the EU is Springer Nature Customer Service Centre GmbH, Europaplatz 3, 69115 Heidelberg, Germany. If you have any concerns regarding our products, please contact ProductSafety@springernature.com

Printed and bound by CPI Group (UK) Ltd, Croydon, CR0 4YY
30/07/2026
02182019-0016